Facing the Future

Facing the Future

The Case for Science

Michael Allaby

BLOOMSBURY

First published in 1995 by
Bloomsbury Publishing plc
2 Soho Square
London, W1V 6HB

Copyright © by Michael Allaby 1995

The moral right of the author has been asserted

A copy of the CIP entry for this book is available from the British Library

ISBN 0 7475 2066 6

10–9–8–7–6–5–4–3–2–1

Typeset by Hewer Text Composition Services, Edinburgh
Printed in Britain by Clays Ltd, St Ives plc

Contents

CONTENTS

Introduction

Our anti-scientific age

We are often told, mainly by non-scientists, that we live in a 'scientific age'. It is usually meant as a complaint, but in any case it is not true. In many parts of the western world, and especially in Britain and the United States, the ideas that dominate our culture are not merely unscientific, they are often anti-scientific, advanced by those who would have the scientific enterprise controlled by non-scientists in order to curtail it. The purpose of this book is to explore the nature and extent of the anti-scientific bias in our culture, to challenge it and to propose a more rational and wholesome alternative.

There is much more at stake here than job security for research scientists. The enemies of scientific research oppose an activity that is rooted in the way we have defined ourselves and our universe for more than two thousand years. To deny those roots is to deny what most clearly characterizes us; we not only abandon our past, we forfeit the future as well.

Warnings of the future cause fear, and a society that fears its own future is decadent. For, if we cannot believe it possible that our children and grandchildren will live happier, healthier, more fulfilling lives than our own we are left with no long-term goals, nothing worth striving for. We may as well make the most of the present, living for immediate material gain and the short-term pursuit of pleasure. The value placed upon wealth, conspicuous consumption and the enjoyment of entertainments that may include portrayals of extreme violence is symptomatic of one that is avoiding the future, of a decadent society.

It need not be so. It is possible to reject the deep pessimism of those who pose the problem in terms that make it insoluble. We might ask, for example, whether the problem really exists at all, or whether the relevant questions have been put, and have been framed correctly, or whether risks are not being exaggerated. We might even question whether it is possible to foretell the future at all. If predictions are, at best, no more than informed guesses, then why should an optimistic scenario be any less valid than a pessimistic one? We could refuse to be influenced by those who believe that the universe is irrational, subject to whimsical intervention by unseen and unidentifiable 'spirit forces', or that reality exists only within our own heads and so everything is subjective. We could restore our faith in progress towards a better society and a better world.

The aim of this book is to redress the imbalance, by challenging the anti-scientific thesis, correcting some misconceptions about scientists and how they work, and suggesting ways in which people might adopt more positive views.

Crossing the boundaries between scientific disciplines, as this book does, allows non-scientists to delight in satisfying their curiosity. The answers to the questions that intrigue us increase our sense of wonder about the world and the beauty of it. It is the blindness to this wonder and beauty, and the lack of curiosity it betokens, that is so pitiable in those who close their minds to the scientific enterprise. There is much excitement, and sheer fun, to be found in the discoveries of scientists, and which contrasts with, and gives the lie to the sterility and sometimes the dishonesty of the anti-scientific position. This excitement will also allow us to reclaim the idea of a future that can be built on to provide a better life for those who follow us.

I

THE ONE-EYED MAN

1

Scientists in Fiction

Paul Ehrlich (1854–1915) was born in Strehlin in Silesia, Prussia (now Strzelin in Poland). As a boy, he hated examinations. Nevertheless he entered university to study medicine, but found himself attracted to chemistry. Largely self-taught, he showed great aptitude for the subject and, after graduating in medicine in 1878, he became a chemist. First he worked as a laboratory assistant and then ran his own small laboratory, before moving on to positions in larger institutions. In 1898 he entered the Institute for Experimental Therapy, in Frankfurt-am-Main.

Throughout his professional life, Ehrlich was motivated by the quest for knowledge. This led him to develop improved laboratory techniques which had widespread application. He found ways to stain cells and tissues for microscopic study, and his staining of the tubercle bacillus provided a reliable method for diagnosing tuberculosis. Ehrlich himself had suffered a mild attack of the disease in 1886, while working on the bacillus in the laboratory of the bacteriologist Robert Koch (1843–1910), and spent some time in Egypt, believing, correctly as it happened, that the dry climate would cure him. From tuberculosis he moved on to collaborate with another bacteriologist, Emil Adolf von Behring (1854–1917), in a study of the bacillus that causes diphtheria, developing an effective treatment for the disease using an antitoxin. This won Ehrlich a professorship at the University of Berlin and Behring a Nobel Prize.

His work on cell and tissue stains suggested to him that, since some of them are taken up by bacteria but not by other cells, there must be some ingredient in the cell to which the stain binds. This being so, it should be feasible to use the stain to kill the cell and, therefore, to produce a drug that would move harmlessly through the human body until it encountered a pathogenic bacterium, which it would then destroy. Such a drug would be a 'magic bullet' that would seek out and kill bacterial parasites. Together with his assistants and colleagues, he set out to construct the 'magic bullet', trying more than nine hundred different compounds. One stain, trypan red, was found to be effective against *Trypanosoma equinum*, the organism that causes sleeping sickness. Its use marked the beginning of chemotherapy, a name that Ehrlich invented. Still greater success came with a compound containing arsenic. Ehrlich discovered it in 1907 as his compound 606, but abandoned work on it because it had no effect on trypanosomes. In 1909 a new assistant found it was effective against the spirochaetes which cause syphilis. Ehrlich called the substance salvarsan (it is now known as arsphenamine). For his work on immunity and serum therapy, in 1908 he shared (with the Russian-French bacteriologist Ilya Ilyich Mechnikov) the Nobel Prize for Physiology or Medicine.

Paul Ehrlich was a hero compelled to work against the grain of his times. As a Jew, he was a member of a persecuted minority. Hitler dismissed his cure for syphilis on the ground that it had been discovered by a Jew and was therefore worthless. He was human in his dislike of examinations. He sought knowledge with unwavering zeal. He was sure of himself and his destiny (indeed, he was extremely quarrelsome and fell out with most of his colleagues at one time or another). The 'magic bullet' was never discovered, so he assumes a tragic quality, but despite this he brought great benefit to humanity.

Scientists in the movies

In 1939 Warner Brothers made a film biography of Ehrlich, called *Dr Ehrlich's Magic Bullet*, which was released the following year. The screenplay was rewritten by John Huston, who had to deal carefully with the subject of syphilis and with Ehrlich's Jewishness. This was referred to explicitly only in the deathbed scene where Ehrlich spoke of Moses, and his quotation from the Pentateuch was cut from the final version. The film starred Edward G. Robinson, who saw in it an opportunity to free himself from the gangster roles for which he was best known. He met members of the Ehrlich family and discovered that he and Ehrlich had many similar character traits and were not too dissimilar physically. In his film make-up, Robinson looked very like Ehrlich. When filming was completed, he said: 'Playing Ehrlich was the easiest acting job I ever did because he was so simple, a great figure struggling with terrific problems'.[1]

If we get the scientists we deserve, it is partly because we invent them to suit our purposes. Paul Ehrlich was presented as a heroic figure on the screen at a time when such figures were needed. War was beginning. Indeed,

Robinson and his wife visited Europe in 1939, discovered a continent in turmoil and, finding there were no tickets awaiting them in Le Havre for the passage they thought they had booked home, managed to leave only because the captain of the *America* gave up his own cabin to them and also one of a nurse for their son and his governess. The ship on which they had hoped to sail but which left France without them, the *Athenia*, was sunk by a German submarine.[2] At that time people needed to be given reason to hope, reason to believe in the feasibility of building a better world, and Ehrlich was recruited to the task. His was the archetypal struggle of good against evil, light against darkness.

In the years that followed, there were other scientist-heroes who received similar treatment, though Edward G. Robinson did not get to play them. They were used as propaganda to sustain public morale by, for instance, releasing heavily censored and romanticized stories about military innovations. The leading characters were portrayed as rather other-worldly 'boffins' who pursued their dreams in the face of opposition from unimaginative bureaucrats and senior military 'brass', presented as ludicrously exaggerated figures of authority. In a film about Reg Mitchell, designer of the Supermarine Spitfire, possibly the most famous of all Second World War fighter aircraft, our hero is seen dozing in his garden when he notices a cloud the shape of Britain drifting across the gentle, pre-war summer sky (his aircraft first flew in about 1935). Other heroes lived dangerously. In *The Small Back Room*, David Farrar played a scientist who specialized in explosives and personally tested the methods he devised for defusing unexploded bombs.

Later, the absent-minded boffin provided an alternative stereotype. *No Highway*, the film of the 1948 novel by Nevil Shute, himself an aviation engineer and at one time managing director of an aircraft factory, dealt with the tragic consequences of metal fatigue in aircraft. James Stewart played the scientist who warned of the danger, but was so absorbed in his calculations as to be endearingly incompetent at ordinary, everyday tasks. In *The Man in the White Suit*, an amiable and innocent chemist, played by Alec Guinness, invented a new polymer from which cloth could be made that would repel dirt and never wear out. The implications of the white suit were not altogether benign, however. The introduction of an everlasting fabric would lead to heavy redundancies in the textile industries, so the invention was unpopular among workers. Eventually, the matter was resolved by having the suit disintegrate. Jobs were saved and we could breathe easily again, saved from perfection.

The Man in the White Suit was a comedy whose central character became tragic as he was reviled and his invention failed. Already doubts were developing about the heroic role in which scientists had been cast. In *No Highway* the amiable boffin prevents a catastrophe by raising the undercarriage of an airliner on the ground, wrecking it to prevent it taking off and, as he had calculated, losing its tail section in flight. The scientist saved the day, but the disaster he averted was one that other scientists, who had ignored his warnings, should have foreseen. The aircraft that had been said to be safe were not so safe after all.

A dark view of scientific research was beginning to emerge, but the scientist-as-idiot had obvious potential and provided material for many talented comedians. This character is fairly harmless, causing explosions and mechanical disasters that do no worse than blacken faces or singe the eyebrows. (Curiously it is always 'he'; when women scientists are portrayed, which is not often, they are cast in heroic mould as, for example, in the film biography of Marie Curie.)

Honey I Shrunk the Kids, one of the recent additions to the genre, took this portrayal a stage further. Its central character is a frenetic inventor who devises a machine for shrinking things. Entering his workshop at the wrong moment, his children are reduced to minute size. The rest of the film deals with their adventures in confronting or avoiding spiders and other beasts larger than themselves and their distraught parents' search for their children. The miniaturization idea has a venerable history in science fiction. H.G. Wells used it in reverse, in *The Food of the Gods* (1904), a story of non-humans grown immense through exposure to a 'miracle fertilizer'. Miniaturized humans featured in the film *The Incredible Shrinking Man* (1957), from a story by Richard Matheson, but most famously in the later film, *Fantastic Voyage* (1966).

The aim of *Fantastic Voyage* was not to amaze or terrify the audience, but to educate them. The story, which began as a movie and was then written as a novel by Isaac Asimov, used miniaturization as a device to allow the cells and other material in the blood and lymphatic systems to be described entertainingly and informatively. Asimov himself pointed out that a mouse-sized man would have a mouse-sized brain and it is hard to see how a man the size of a single cell would have room for any brain at all.[3] Such ideas are, of course, preposterous but, the Asimov story apart, they are usually employed, as in *Honey I Shrunk the Kids*, to show scientists as irresponsible fools whose meddling in matters they do not understand can bring catastrophic results. True, the central characters usually manage to return to their proper size, and in *Honey I Shrunk the Kids* the crazy scientist rescues his children, but the message is clear:

scientists may be affable, their antics may be funny, but we would be wise not to trust them.

This is not how other professions are presented in novels and movies. For example, in portrayals of poets, painters, musicians, writers or actors and the psychological traits that draw them to their respective professions, the struggles that those entail, and the effect on the artist's character, no one questions whether these occupations are worthy in themselves. It is assumed that they are. Doctors, too, are usually presented in a positive light: they relieve human suffering, so their motives must be good. Only scientists are treated less than even-handedly. The value of their work is not taken for granted; it may be beneficial, but it may also be malign.

Perhaps scientists should feel flattered that people pay attention to the social implications of their work. If so, should artists, by the same criterion, be concerned that their work seems to be taken less seriously; or theologians and philosophers worry that they are pretty much ignored, unless we are to suppose that their ideas are too difficult, and unsuitable material for films or literature?

A change in attitude

All fame carries a price, for the famous exist in some sense as public property, to be used or abused according to the fashion of the moment. If scientists have benefited from the attention of writers and film makers, they have also suffered, because it is from novels and even more from movies that most people receive their values. To some extent, every film biography caricatures its central character. The story must be made interesting and emotionally and visually appealing. Anything that in real life is repetitious, humdrum or routine is played down or condensed in favour of the highs and lows of triumph and despair. Events may be rearranged or even invented to heighten the drama. All this is standard procedure, but for scientists there is more. Their work is judged at the same time as it is portrayed and new times bring new judgements.

The change in attitude began after 1945. In wartime, scientists were honoured for devising new weapons or protective measures against them. They played a distinguished, and therefore featured role in the defence not only of their country but of the liberal democratic philosophy on which its institutions were founded. By the end of the war, the weapons that had found no targets remained in an uncertain world. As reassurance at victory gave way to peacetime unease, the weapons began to appear threatening and the protection against them inadequate. Nuclear weapons in particular struck fear into the hearts of a whole generation. In both eastern and western power blocs the weapons were under strict military control and the military under strict political control. They could not be used without political sanction, but it was not politicians who were blamed for their existence so much as the scientists whose findings had made their development possible and, therefore, inevitable as each side strove to take a lead in the arms race. According to Hilary and Steven Rose,

> Before the Second World War scientists generally appeared in novels, children's stories and so forth as rather endearing, absent-minded figures, possibly mad, preferably possessed of pretty daughters, but on the whole fairly innocuous, inventors of machines that did easy things in a complex way but which often failed to work. This is far from true today. The scientist, in novel, play, film and comic, is a figure of power, sometimes sinister, sometimes naive and virtuous but if so a helpless tool in the hands of those who wish to misuse him. The machines he invents, rockets, atomizers, death rays, are dangerous and of immense military potential . . . above all, he [the scientist] is no longer a figure of innocent fun. No one remains innocent about the potential application of seemingly highly theoretical research.[4]

Certainly, scientists should not be immune from criticism or their work hidden from public scrutiny. Public curiosity about their activities is entirely healthy and goes beyond wishing to map the benefits of new gadgets derived from their researches. Scientific findings affect the way we see the world about us, helping us to understand and adjust to the picture they reveal. We can participate only if we are well and accurately informed. The problem is not one of lack of scrutiny, but of uninformed scrutiny that is influenced by the negative stereotypes found in popular entertainments. Over the last half-century we have swung from the heroics of Dr Ehrlich's Magic Bullet to the disdain of Honey I Shrunk the Kids, and there is worse, for in other stories, some of them not so new, the work of scientists is described in even less flattering terms. The 1993 film Lorenzo's Oil, for example, portrays medical scientists as largely incompetent. Augusto and Michaela Odone acquire the scientific knowledge they need to devise a cure for the adrenoleukodystrophy (ALD) from which their son, Lorenzo, suffers. ALD is a fatal illness, genetically inherited through the X chromosome, and the cure the Odones discover is erucic acid triglyceride. Unfortunately, there is doubt about the efficacy of 'Lorenzo's oil' and the side-effects, so the search continues for a, probably more appropriate, gene therapy. One reviewer described the film as 'entrancing', but 'at base, pernicious'.[5]

In part, the story reflects the public's attitude and the commercial desire to tell people what they are prepared to hear. Admittedly, this is primarily popular and usually hugely enjoyable entertainment, but there is another side to the story that responsible writers and film makers should perhaps bear in mind. The scientific enterprise is intimately linked with our belief in and hope for the future, our idea of progress. If the portrayal of scientists as inherently dangerous figures damages that belief and encourages us to reject a rational basis for examining the world, then it is one-sided, irresponsible and harmful.

Buffoonery may redeem the meddling scientist, but not all fictional meddlers are clowns. Thwarted ambition and confusion provide laughs, but often the laughter conceals the allegation of hubris. Strip away the humour and the overweening pride, and lust for power, or sometimes greed, are exposed. These are not funny and are not meant to be. They serve as a warning to those who presume to interfere with nature or creation, who think they have a right to challenge the universe, that they do so at their peril. They therefore and, by implication, all scientists working at the frontiers of research are a danger to us all.

Fictional monsters

The most famous of all fictional scientists obsessed and eventually destroyed by such a challenge is, of course, the eponymous Frankenstein of Mary Shelley's novel. The story has been told so often, in print, comics, and movies, with so many alterations, and it has served as the archetype for so many other stories, that the original risks being forgotten.

The novel begins with a ship on a voyage of exploration rescuing Victor Frankenstein from a sledge marooned on a large ice floe. He is exhausted and close to death. As he recovers, he tells his story to the leader of the expedition, who describes him as 'a man who, before his spirit had been broken by misery, I should have been happy to have possessed as the brother of my heart'.[6] He recounts how, as a young man from Geneva, he attended the university at Ingolstadt. There he met a professor who informed him that modern natural philosophers (scientists) 'ascend into the heavens; they have discovered how the blood circulates, and the nature of the air we breathe. They have acquired new and almost unlimited powers; they can command the thunders of heaven, mimic the earthquake, and even mock the invisible world with its own shadows.'[7] Inspired, Victor embarked on his own researches until, 'After days and nights of incredible labour and fatigue, I succeeded in discovering the cause of generation and life; nay, more, I became myself capable of bestowing animation upon lifeless matter.'[8] Having gathered together bits and pieces of corpses, he assembled them into a complete body, of 'gigantic stature' because he found it easier and quicker to work with larger components, until, 'With an anxiety that almost amounted to agony, I collected the instruments of life around me, that I might infuse a spark of being into the lifeless thing that lay at my feet.'[9] But when the monster came to life he fled from it. Gentle and loving, but hideous to behold, the monster was driven by the humans he encounters into increasingly embittered solitude, from which Frankenstein refused to rescue him by making a female companion for him. The rest of the story describes the tragedy of their entwined lives.

The circumstances under which Frankenstein came to be written are well documented. In the summer of 1816, Mary Shelley and her husband, the poet Percy Bysshe Shelley, accompanied by Mary's stepsister, Claire Clairmont, went to Switzerland for a holiday, staying near to Geneva, where they were neighbours of Lord Byron. That year the summer weather was dismal, for reasons which are of some interest to scientists. The greatest volcanic eruption of modern times occurred in 1815, when Mount Tambora, in Indonesia, injected large amounts of particulate matter into the upper atmosphere. The particles reflected incoming solar radiation, producing cool surface conditions.' It proved a wet, ungenial summer,' Mary Shelley wrote, 'and incessant rain often confined us for days to the house.'[10] To pass the time the friends read ghost stories and then Byron suggested each of the four of them should write one. The poets tired of the exercise and only Mary succeeded. Mary recounted how Byron and Shelley had been holding long philosophical conversations in the course of which they discussed the nature of 'the principle of life':

> They talked of the experiments of Dr Darwin (I speak not of what the doctor really did, or said that he did, but as more to my purpose, of what was then spoken of as having been done by him), who preserved a piece of vermicelli in a glass case till by some extraordinary means it began to move with voluntary motion. Not thus, after all, would life be given. Perhaps a

corpse would be reanimated; galvanism had given token of such things; perhaps the component parts of a creature might be manufactured, brought together, and endued with vital warmth.[11]

The identity of 'Dr Darwin' is uncertain: Charles Darwin was only seven years old in the summer of 1816 and in 1831, when Mary wrote this account, he had just set sail on HMS *Beagle*; his father, Robert, was a physician but not a scientist, and his grandfather, Erasmus, who was an eminent scientist as well as physician, had died in 1802.

The 'vermicelli' story seems to refer to the idea of spontaneous generation, according to which living organisms might arise spontaneously from non-living matter. The view had been widely held since antiquity, but was being strongly challenged by the eighteenth century; efficient microscopes allowed biologists to observe the reproduction of very small organisms. In those days, the word 'evolution' had a different meaning from the one it has today. It implied a revealing of immanent qualities, like the development of a plant from its seed, and was combined with the idea that all living things comprise a 'great chain of being', the lowest at the bottom and Caucasian humans at the top. The 'power of life' drove the increasing complexity by means of which organisms could ascend the chain. It followed from this that the chain was constantly growing and, therefore, new organisms must constantly be added at the base, which is where spontaneous generation occurred. The most distinguished proponent of this view was Jean Lamarck (1744–1829). Erasmus Darwin had written in support of it and the holidaying party might well have discussed it.

'Galvanism' may also have been a topic of conversation. Twenty-five years had passed since Luigi Galvani (1737–98) had published his theory of 'animal electricity' based on experiments in which frog muscles twitched when brought into contact with iron and copper simultaneously. What they could not have known was that, some years later, the Scottish publisher Robert Chambers (1802–71) would make much of an experiment that claimed to have generated life by passing an electric current through a mixture of potassium silicate and copper nitrate solutions. It transpired that a common mite had entered the solution unobserved.[12] Most of the movies had Frankenstein use the passage of an electric current to animate his creature, sometimes by harnessing lightning from a storm introduced to add a suitable hint of the supernatural. The original story does not describe this, though it does tell us that the event occurred on a dismal, rainy night and, in her introduction, Mary Shelley mentions 'the working of some powerful engine'.

11

ught together items that were probably familiar to intellectuals of her
to produce the most terrifying story she could imagine. 'Frightful it
st be;' she wrote in her introduction, 'for supremely frightful would be
he effect of any human endeavour to mock the stupendous mechanism of the
Creator of the world.'

There we have the ingredients and the motivation, not only of the original
story, but of all those cast in its mould. It begins with popular scientific
misconceptions, garbled accounts that were magnified in the course of
conversation by people who had not read the publications on which the
ideas they were discussing were based and who lacked the scientific
understanding to evaluate them. As reported, these accounts seemed to
represent a challenge to 'the Creator of the world', a challenge to accepted
beliefs. Seen in this light they were inevitably doomed.

We have lived with *Frankenstein* for more than a century and a half but it
would be a mistake to think the theme is exhausted. Its latest incarnation,
albeit with a very important modification, is to be found in the novel, and
movie of the same name, *Jurassic Park*.[13] Here, it is not a scientist who
challenges 'the Creator of the world', but a millionaire showman, a man who
started his career running a flea circus. The scientists who bring his dream to
life are hirelings. As ambitious as any scientist of the genre, and as careless of
the consequences of their work, they lack the moral stature of Victor
Frankenstein. However, Michael Crichton, the author, and Steven Spielberg
who made the movie, are too well informed and too sympathetic to scientific
endeavour to cast scientists in the roles of villains. Other scientists warn of the
danger and try to redeem the situation when it has got out of hand and the
dinosaurs run amok. These animals were undoubtedly large and well armed,
as they needed to be in the environment they inhabited, but the book and
film make them vicious and insatiable as well. This transforms fascinating
animals which existed historically into nightmare monsters, and their
terrifyingly destructive orgy is a warning to those who presume to tamper
with the natural order of things.

Although a splendid yarn, it still contains other classic 'Frankenstein'
ingredients. It is not possible to regenerate dinosaurs in the way proposed, or
indeed at all. Genetic material has been obtained from insects preserved in
amber, but whether DNA could be extracted from blood consumed by such
an insect is extremely doubtful. If DNA were obtained, it would represent
only a minute fraction of the complete genome of the original animal and
filling the gaps with DNA from an unrelated organism simply would not
work. Even supposing a complete genome could be obtained, to grow into an

animal it would need an appropriate environment, which in this case would be a fresh dinosaur egg.[14]

Some of the dinosaurs are portrayed realistically in the movie, but not all of them. That is to say, real animals would not behave as the carnivorous species in the film do, though the herbivores may well have closely resembled those on the screen. Carnivores kill only in order to eat. They do not spend all their time hunting and killing for the sheer fun of it. After each meal, the animal would spend a long time digesting, reacting to the presence of prey species only if it were disturbed. Caribou will ignore a wolf walking through the middle of the herd if they see the wolf is not hungry. If carnivorous dinosaurs were ectothermic, like crocodiles and snakes, they might well have spent days or even weeks digesting a meal. If they were endothermic, like lions, they would have spent time between meals just lying around or dozing. This is not a matter of temperament, but of energy conservation. Our knowledge of dinosaur behaviour is mainly speculative, of course, but there is no reason to suppose it diverged so widely from that of mammals filling similar ecological niches as to defy basic metabolic principles, which the insatiably voracious *Tyrannosaurus rex* and *Velociraptors* of the film certainly do.

The spectre of annihilation

Frankenstein and *Jurassic Park* are fantasies, scary but far removed from the real, everyday world. The danger of the real world can, of course, also be exploited. The film *Dr Strangelove or How I Learned to Love the Bomb* (1963) was a black comedy about the possibility of a full-scale nuclear war. It was funny, but not fantastic. The 'doomsday bomb' referred to in the film, which relied for its effect on fallout rather than immediate blast and heat, might be feasible, though it would consist of a large number of bombs detonated in carefully designated locations rather than a single weapon.[15]

The film was mainly a satire on the doctrine of Mutual Assured Destruction, better known by its acronym of MAD. Some critics have suggested the figure of Dr Strangelove was based on the physicist and mathematician Hermann Kahn (1922–83), director of the Hudson Institute, in the United States, where the concept was devised. It was Kahn who claimed to 'think about the unthinkable' by speculating on the consequences of a real thermonuclear war. MAD was adopted early in the presidency of John Kennedy and was finally abandoned by President Carter in 1980.[16] The acronym contributes nothing to the credibility of the idea, but it was the product of an arms race in which the United States and Soviet Union found themselves so well matched that the prospect of surviving an initial attack on military targets long enough to retaliate appeared impractical. Weapons were, therefore, to be targeted at cities, so that any attack launched by either side would trigger a retaliatory strike against the main cities of the aggressor. Thus the initiation of a war would guarantee the destruction of both combatants. The strategy was grim, but it was intended only to deter. Some might argue that it proved successful for nearly twenty years.

As for Hermann Kahn, in 1976 he was still warning of the dangers of thermonuclear war. 'Fortunately the present leadership of each superpower is aware of the enormous destruction such an exchange would cause,' he wrote, 'and this awareness in itself exercises a very strong restraint. However, even a very small nuclear war could do an extraordinary amount of damage'.[17] He goes on to outline ways in which the problem might be addressed and the risk of conflict reduced. These are hardly the words of a warmonger; nowhere does he advocate the destruction of an 'evil empire'. That idea came much later and by then Kahn was dead. He was an optimist who believed strongly in the possibility of progress based on the proper application of scientific understanding to economic and social problems.

Dr Strangelove is an another example of an entertainment in which a chain of events leading to disaster is based on the exaggeration or misinterpretation of scientifically inspired ideas. The scientist responsible for them is portrayed as a deluded megalomaniac, incapable of noticing even the simplest and most obvious dangers inherent in his scheme.

But does this really matter? After all, such stories are intended to amuse not to instruct. We do not expect to acquire our understanding of scientific concepts from them, but from informed people who make serious attempts to explain the ideas to us. We learn from our schoolteachers and from books, newspaper and magazine articles, films, and radio and television programmes.

This argument is true only up to a point. Otherwise, we might expect more stories where scientists are shown in a positive light, promoting schemes that do not contradict basic scientific principles. Some stories are like this, and they can be very exciting, but they are few. Another argument is that the dramatic theme of hubris leading to disaster is not restricted to scientists but recurs throughout our history and has always been popular. This is also true, but it overlooks the purpose for which the theme is used: to warn humans against assuming god-like powers and challenging the universe. The emphasis on scientists in modern interpretations of the theme strongly suggests that they are the ones who are most likely to be guilty of hubris. In the real world, politicians experimenting with social engineering, economists imposing their theoretical systems on entire nations, and lobbyists unashamedly distorting facts in a bid to gain popular support are also guilty of hubris, but they are seldom treated as unkindly by fiction as scientists (perhaps because they are more likely to fight back).

It is often said that artists and writers depict the world they see about them with a kind of objectivity, a 'special eye', placing them outside what they describe and giving them a sense of detachment. This is nonsense. Artists cannot, any more than the rest of us, escape the influence of the opinions and attitudes of a society of which they are part. Their work reflects their personal view, but it is a view shared by society at large or by the circles in which they move. A negative representation of a scientist in a work of fiction reflects a much more widely held opinion than that of the authors' alone.

This is the point. Our stories reflect our own ignorance and fear. They betray the failure of nerve that makes us unwilling to contemplate the future because we cannot imagine a future that is not worse than the present. We are the victims of our self-inflicted cultural degeneration and scientists are our scapegoats.

2

Scientists, the Barbarians at the Gates

Once upon a time many people believed the Earth was flat, like a disc. Presumably they felt comfortable with this image, which conformed to the evidence of their own eyes. Eventually, though, they were forced to abandon it because scientists had demonstrated beyond the remotest possibility of doubt that our planet is spherical, but slightly flattened at the poles. Some people believed that the Earth rests upon the back of an elephant, itself supported by a turtle. The picture is attractive, but it turned out to be untrue.

In every culture people have devised their own explanations for the way the world and all living beings came into existence. The accounts differed widely and contradicted each other, and in the end all of them turned out to be wrong. Close observation and calculation led to the formulation of alternative ideas which built into a partial picture of the universe consistent enough to include a wide range of phenomena. So it goes. The accumulation of knowledge leads to better explanations and old ideas are challenged and then overturned. It is no wonder that scientists are distrusted. They are profoundly subversive.

Someone holding a view of the universe which is radically different from that held by most people may well experience difficulties in communication. Each of the opposing views might generate its own imagery, so metaphors and similes are used in different ways, leading to misunderstanding. Variations in the concepts around which thoughts are constructed might give further cause for confusion. The 'radical' might be thought of as speaking differently from the way most people speak, almost like a foreigner, which is one meaning of the word 'barbarian'. 'Barbarian' is also a name applied since the sixteenth century to non-Christians. In both these senses, of speaking strangely and crudely and of subscribing to non-Christian beliefs, scientists are regarded by some people as barbarians.

Science and religion

Many people, scientists and non-scientists alike, suppose that the scientific view of the world has supplanted the religious view from which it arose. It was, after all, the Christian belief in a universe created on consistently rational principles that invited systematic investigation and logical conjecture. An irrational universe, subject to the ever-changing whims of manipulative spirits, could not be studied at all. 'Science, like religion, provides a meaning that connects all it touches,' writes Bryan Appleyard.[1] 'But it is a more limited meaning that offers a form of truth without significance.' Appleyard blames scientists for the rise of the liberal views he opposes, but the charge of atheism is a common one, and to many people atheism is an offence.

Offence or not, we live in a society which numbers religious believers of many kinds, agnostics and atheists among its members. The scientific community, as a subset of society at large, also includes people holding these views. Probably, the scientific community contains a higher proportion of atheists than do non-scientific groups, but this may be due in part to the fact that scientists are compelled, by virtue of the intellectual disciplines they practise, to think more deeply than most people about religious issues. What passes for religious belief in our society often consists largely of an acceptance of a kind of 'folk religion' that makes no intellectual demands and could not long withstand rigorous examination.

Historically, however, most scientists have held religious views. Many have been devoutly religious and not a few of them were priests. While undoubtedly true, this statement is somewhat misleading, at least in England, because until a little more than a century ago the Universities of Oxford and Cambridge would not award a degree to a dissenter, and teaching posts were usually held by ordained clergy. There may well have been individuals who concealed their disbelief rather than risk blighting their careers, though there is no reason to suppose the great majority were insincere in their profession of faith. Gilbert White (1720–93), famous as the author of *The Natural History of Selborne*, was a curate, the chemist Joseph Priestley (1733–1804) was a Unitarian minister, and Michael Faraday (1791–1867) and James Clerk Maxwell (1831–79) were both deeply religious, though they often challenged accepted views, including religious ones. Priestley declared that the spread of scientific knowledge would be a means of 'putting an end to all undue and usurped authority in the business of religion.'[2] It is especially odd that Appleyard should cite both Maxwell and Faraday in his muddled indictment of 'the real ambitions of our scientist–priests'.[3]

Scientists who were Christians were the rule and sometimes their religious views coloured their attitude to their work and supplied some of the imagery they used in describing it. 'Would the notion of Maxwell's Demon have occurred to somebody with a different upbringing?' the philosopher Mary Midgley asks. 'The original forging of the modern understanding of electricity owed nothing to atheism.'[4] Maxwell was Cavendish Professor of Experimental Physics at Cambridge University and set up the laboratory of that name there, in 1874. Over the gateway he had inscribed: 'The works of the Lord are great, sought out of all them that have pleasure therein' (Psalm 111, verse 2). When the Laboratory was moved some years ago, those words were written up to greet visitors to the new premises. Maxwell was certainly no atheist and his religious views continue to command the respect of his successors.

Maxwell, Faraday and countless other scientists who were devoutly religious saw no contradiction, far less blasphemy, in contributing to the revelation of the structure and function of the universe and all it contains. The scientific route to this revelation is as valid as any other, but more effective. This is how the famous passage with which Stephen Hawking ends his *A Brief History of Time* should be understood. After describing the search for a unified theory that would link the four physical forces and so bring all the laws of physics within a single framework, Professor Hawking says that in principle this theory should be intelligible to everyone.[5] 'Then we shall all, philosophers, scientists, and just ordinary people,' he says, 'be able to take part in the discussion of the question of why it is that we and the universe exist. If we find the answer to that, it would be the ultimate triumph of human reason – for then we would know the mind of God.'[6] Of course, this may be nothing more than a flowery use of language appropriate to the final page of a book, but some critics have taken it seriously and denounced Hawking for it. Both Appleyard and Midgley seem to consider it the height of impertinence.[7] Is it wrong to pursue paths that were also followed by Newton, Maxwell, Faraday and all the others? Or is it wrong, if those paths lead to areas of knowledge that may prove controversial, to deny scientists access to them? If there is knowledge to which we are not entitled, where shall the boundaries be drawn, and by whom? Should we allow them to be defined by opponents of a scientific enterprise that they do not fully understand? Or is it not, perhaps, better to know even if the knowledge causes us radically to revise certain of our cherished ideas?

Refusing to accept some of the more conventional religious concepts does not imply an absence of proper religious sensibility. Freeman Dyson (born

1923), a mathematician and theoretical physicist who was born in England and has been a professor of physics at the Institute for Advanced Study in Princeton since 1953, rejects what he regards as the artificial division between science and religion.[8] 'Being a scientist, trained in the habits of thought and language of the twentieth century rather than the eighteenth,' he writes, 'I do not claim that the architecture of the universe proves the existence of God. I claim only that the architecture of the universe is consistent with the hypothesis that mind plays an essential role in its functioning.'[9] Dyson is profoundly optimistic and his 'religious' views were formulated in response to a statement by fellow physicist Steven Weinberg that, 'the more the universe seems comprehensible, the more it also seems pointless'. 'No universe with intelligence is pointless,' Dyson retorted. 'No matter how far we go into the future, there will always be new things happening, new information coming in, new worlds to explore, a constantly expanding domain of life, consciousness and memory.'[10]

It is still possible for a scientist to hold more orthodox religious views and many do. John Polkinghorne, for example, is a physicist, a Fellow of the Royal Society of London and an ordained minister of the Church of England. In a lecture to the Royal Society of Arts in October 1992, he outlined the reconciliation he has found between his scientific and religious views. Rejecting all ideas of God as 'a kind of capricious celestial conjuror' or as one who 'interacts with the world by scrabbling around at its subatomic roots', he used the principle of indeterminacy and chaos theory to argue that, 'the physical world, even at the everyday level, is something more subtle and supple than a merely mechanical universe.' Moreover, he said, 'we can take with all seriousness all that science tells us about the workings of the world and still believe that the God who holds it in being has not left himself so impotent that he cannot continuously and consistently interact within cosmic history. Such a God is one to whom a scientist can pray with complete integrity.' Near the end of his lecture, he said:

> Many people today find it difficult to pray because they find it difficult to believe in a God who does particular things in particular circumstances. Part of their reluctance stems from a feeling that science has disposed of that possibility by its exhibition of a physical world whose history is the inexorable outworking of deterministic and impersonal laws. But such a world would not only be one in which God did not act but also one in which we could not act either. I hope I have persuaded you that if we take

modern science absolutely seriously we shall not find ourselves condemned to so rigid and implausible an account of what is going on.[11]

Here, then, is the possibility of reconciling two supposedly conflicting views. It rests on the rejection of two false beliefs: the belief in God as a magician and religion as magic; and the belief in a mechanistic, deterministic universe. The first belief is unacceptable to modern Christian theologians and the second to many scientists, and especially to physicists, though determinism remains a problem. Unfortunately, it does not resolve the conflict, because too many people find it impossible to abandon ideas that are dear to them. 'The world chose science rather than magic', writes Appleyard of the diversity of Newton's prolific output, 'because it thought it was "true", because it worked. What must never be forgotten, however, is that it *was* a choice, we adopted a particular perspective.'[12] In other words, either alternative might have suited as well. This is, of course, preposterous. Scientist or sorcerer, you may suffer if, say, a fungal disease destroys the harvest, indicating that for all practical purposes there is a real world outside our own heads for which the scientist, but not the sorcerer, can offer an explanation that may help prevent such a disaster in the future. A choice between demonstrable reality and fantasy is hardly a choice at all.

Scientists, poetry and culture

Scientists also err in sometimes speaking or writing poetically and in making jokes. Midgley fulminates against the 'Omega Point', the point at which life will have spread throughout the universe and will have stored details of everything it is possible to know. The idea was suggested in 1986 in The Anthropic Cosmological Principle, by John D. Barrow and Frank J. Tipler. In its 'weak' form, the anthropic principle says that intelligent life can develop only in certain parts of the universe and at times when conditions are favourable, so we should not think it remarkable that our corner of the universe is hospitable to us. We have no knowledge of possible regions where it is not. The principle also explains why the 'big bang' happened when it did: it takes that long for intelligent beings to evolve. In its 'strong' form the principle goes further. It says because the universe – or our universe since there may be others – is the way it is, intelligent life was bound to emerge within it and it is the presence of observers that gives the universe a tangible reality. The 'weak' version is widely accepted, but few scientists take the 'strong' version very seriously. Barrow and Tipler also suggested that the final collapse of the universe would be prevented by shovelling matter into black holes. If both ideas are not jokes, they are speculation, thoughts worth discussing but not for a moment intended as predictions. Should scientists never speculate or jest for fear of being misunderstood?

Misunderstandings that arise from too literal an interpretation of the written word can also affect non-scientists who write on scientific matters. A few years ago I wrote a popular account of the 'Gaia' hypothesis.[13] Seeking some purpose for humans that seemed consistent with the theory, I suggested that if, as had been proposed, 'Gaia' were a single, living being, then humans might be its reproductive mechanism. You can, indeed, make some kind of case for this, though I meant it as a joke. Later, I learned that I had been taken seriously by certain 'New Agers', who had come to see their lives in an entirely fresh light.

Not surprisingly, irreligious scientists are also portrayed as unimaginative and soulless, the last people you might think of as reading poetry, far less writing it. 'The idea that poets "think", that poetry is itself a rigorous, highly disciplined art is quite foreign to them [scientists],' writes Midgley.[14] Really? Sir Humphry Davy (1778–1829) would not have agreed.

> By science calmed, over the peaceful soul,
> Bright with eternal Wisdom's lucid ray,
> Peace, meek of eye, extends her soft control,
> And drives the puny Passions far away.[15]

Sir Francis Bacon also wrote poetry, and so did Gilbert White, James Clerk Maxwell and, more recently, Sir Julian Huxley (1887–1975), not to mention many less well-known names. According to Tim Radford, 'the most famous English poet in 1793 wasn't Wordsworth or Blake, it was a scientist called Erasmus Darwin' (1731–1802).[16] *The Loves of Plants*, Part II of his *The Botanic Garden*, published in 1792, was a bestseller. The quality of the verse shows why:

> With fierce distracted eye IMPATIENS stands,
> Swells her pale cheeks, and brandishes her hands,
> With rage and hate the astonish'd grove alarms,
> And hurls her infants from her frantic arms.

As well as scientist–poets, there were also many poets who were inspired by scientists and their ideas. Byron wrote about the prehistoric reptiles that were

21

later called dinosaurs, Samuel Taylor Coleridge attended Davy's lectures to seek out new ideas, and Shelley went further than any of them. He tried his own experiments and then described them in verse.

Scientists should also beware of attempting fiction, though many do and use it to explore ideas in an informal way. In 1957 Fred Hoyle published *The Black Cloud*, a story about an intelligent being in the form of a cloud of diffuse matter. The cloud threatened the Earth and humans had to find a way of communicating their predicament to it. The theme centred on how we might establish communications with a life-form so vastly different from anything with which we are familiar. The story has been misunderstood, however, as suggesting that this was a physical form intelligent life might take.

Freeman Dyson's autobiography, *Disturbing the Universe*, is frequently punctuated by literary quotations and filled with allusions. These occur even in his chapter titles: 'The Redemption of Faust', 'The Ascent of F6', 'The Island of Doctor Moreau'. It is clearly the work of a man who reads much. He also loves music. He visits colleges to talk to undergraduates, preferring the smaller and more obscure campuses, and beams with pleasure on recalling how on one such visit the early arrival of his flight allowed him to attend an 'absolutely superb' concert of sixteenth-century music.[17]

Most ordinary, rank-and-file scientists lead fairly humdrum lives, like the rest of us. It is the 'stars' who attract attention, the senior professors, Nobel Prize winners and leaders of the teams that achieve breakthroughs that can be sensationalized by the press. However, no one should be surprised to discover that those at the forefront of their profession are often very cultured people, with a love of art, music and literature. In Britain, the Master of the Queen's Music, composer Malcolm Williamson, is also a distinguished research biochemist. Einstein played the violin well but other physicists have also displayed musical talent. The British theoretical physicist Paul Dirac (1902–84), whose love of his work derived from an aesthetic appreciation of the beauty of mathematics,[18] adored music, according to his widow. 'Even my knitting had to stop for complete silence when he was listening. He was also a great admirer of art. Not only did he like beautiful things in our home, he was also a tireless museum fan. He made me read *War and Peace*, and he read a great many books that I suggested to him. Theatre, movies, ballet: we never missed a good performance, even if we had to go to London or from Princeton to New York'.[19]

All of us are products of the same culture, a culture that includes art, music and literature, to which we are exposed throughout our whole lives. The language that we use is taught to us partly through our literature. Scientists, therefore, are as likely as anyone else to read novels and poetry and perhaps to write them, to listen to and possibly perform or even compose music, to appreciate art and perhaps to paint, and to enjoy plays, opera, ballet and movies. Furthermore, it is perfectly possible that individuals who are supremely gifted in a scientific field may also be gifted in others. The practice is different, but the creativity of scientists and artists springs from the same source.

It is unjust to regard scientists, in their capacity as scientists, as soulless barbarians with no appreciation of beauty who are ignorant of the impact their discoveries may have on established beliefs. Most people would agree that beliefs should be abandoned when they cease to serve any useful purpose and that a new system of beliefs should replace them. Scientists alone cannot supply that system, however. They are not empowered to do so and have no wish to impose their beliefs on others. If a new belief system is to emerge, or an old one be rehabilitated, it takes many thoughtful people to participate in the task. Scientists can make a contribution, but the collaboration of non-scientists from many walks of life is vital.

Religion and morality

In 1927 a conference of physicists was held at Solvay, in New York State, and as is usual at conferences, many of the delegates stayed in the same hotel. One evening, a group of the younger scientists, including Werner Heisenberg, Wolfgang Pauli and Paul Dirac, found themselves discussing religion. Einstein, one of them commented, often talked about God, whom he believed to be involved in the immutable laws of nature. Max Planck, on the other hand, took a simpler view, which Heisenberg had pieced together from conversations with people who knew the great man well and which he outlined for his friends. Planck believed science and religion were fundamentally compatible because they dealt with quite distinct facets of reality. He held that science invites its practitioners to make accurate statements about objective reality and to grasp its interconnections, while religion deals with values, with what ought to be or what we ought to do. Thus scientists seek to discover what is true or false, religious thinkers what is good or evil, noble or base. Heisenberg himself could not accept so sharp a distinction between the objective and subjective, between knowledge and faith. Pauli, says Heisenberg, shared his concern:

At the dawn of religion all the knowledge of a particular community fitted into a spiritual framework, based largely on religious values and ideas. The spiritual framework itself had to be within the grasp of the simplest member of the community, even if its parables and images conveyed no more than the vaguest hint as to their underlying values and ideas . . . That is why society is in such danger whenever fresh knowledge threatens to explode the old spiritual forms . . . In western culture, for instance, we may well reach a point in the not too distant future where the parables and images of the old religions will have lost their persuasive force even for the average person; when that happens, I am afraid that all the old ethics will collapse like a house of cards and that unimaginable horrors will be perpetrated.[20]

To Dirac, God was an invention of the human imagination and religion served mainly to sustain the power of the ruling classes.

Scientists have always worried about religion and the dangers of destroying its mythology. Like most people, they accept the need for a system of beliefs to underpin morality. The difficulty arises because belief is essentially irrational. A proposition is not verified by our belief in it, nor disproved by our disbelief. If the validity of a proposition can be demonstrated rationally we are not required to believe or to disbelieve. Provided that we understand the premises and arguments supporting a proposition and accept them as logically consistent, we know that it is true, though proper intellectual rigour may require us to accept it as only provisionally true (see chapters 7 and 12).

The link between systems of religious belief and moral concepts is subtler than is often supposed. Atheists and humanists protest vigorously that it is possible to construct a moral philosophy without reference to religion and, of course, they are perfectly correct. To counter the charge that they derive their own moral philosophies from those developed by theologians and merely discard the religious foundations, there is more persuasive evidence. If morality depended fundamentally on religion, the wide variation in religious beliefs could reasonably be expected to produce a similar variation in moral systems, but this is not the case. Christians, Hindus and Buddhists, for example, differ greatly in their religious opinions, but the codes of moral behaviour that they derive from them are remarkably similar. Otherwise, the governments of pluralist societies would find it almost impossible to pass laws that their peoples found morally acceptable. That they can do so is due not so much to religion as to Utilitarianism, the secular philosophy developed principally by Jeremy Bentham (1748–1832) and John Stuart Mill (1806–73).[21]

Indeed, where religion teaches that behaviour leads to rewards or punishments it may be counter to morality. This is because, to many moralists, a constrained choice, one made in hope of reward or fear of punishment, is morally insignificant. For example, we do not hold an action to be morally reprehensible if it is performed by a person under extreme duress, such as when a gun is held to their head, or morally admirable if it is performed, even partly, for personal gain.

Morality is, in any case, a topic on which modern scientists have something useful to say. It bears a close similarity to altruism, a type of behaviour that has long puzzled ethologists. An altruistic act is one performed by an animal at a cost to itself, even the loss of its life, from which another animal benefits. The classic example is the alarm call with which a bird warns others of the presence of a predator at the risk of attracting attention to itself. There are several possible evolutionary explanations,[22] and one, which seems relevant to human societies, derived from game theory.

This developed from an attempt to find a solution to the 'the prisoner's dilemma', essentially to determine the optimum strategy to pursue in a series of transactions between two participants who may collaborate or defect.[23] You can think of it as a game with an indeterminate number of rounds. Two players each have a box with a lid. One at a time, each player, unseen by the other, places a token in their own box or leaves it empty, then opens the other's box and removes what, if anything, the other player has left in it. Obviously, the player who leaves nothing can lose nothing and gains if the other box contains a token, but a winning strategy for one round fails in the long term. The winning strategy, which defeated all others, was called 'tit-for-tat'. In the first round you cooperate, by leaving a token in the box. In each subsequent round you do whatever your opponent did last time: if they left a token you leave a token, if they left nothing you leave nothing.[24] Since this strategy was first published it has been further refined.[25] Simple though it is, this demonstration shows that only through cooperation and honest dealing can social relationships be sustained indefinitely. To use a jargon term, the strategy is 'stable'. It may strike you as a thin basis for morality, but it is an exceedingly powerful one and may well explain how cooperation evolved and acquired its great value.

Religion is not necessary as a secure foundation for morality. Its more important role is, so to speak, locational. That is, it can explain to us where we stand in relation to one another, to the world we inhabit and to the universe, and from this explanation many people derive a meaning for their lives. Since the traditional mythology in which religious explanations are

25

expressed is no longer credible, we need to discover new statements, new accounts, that accord with scientific knowledge to supply the needed explanation. Many scientists have attempted the task.

Modern physics and the religious quest

In two influential and scholarly books, the physicist Fritjof Capra explored what he believes are parallels between the discoveries of quantum theory and the mysticism of Buddhism, Hinduism, Confucianism and Taoism.[26] At the level of the extremely small, but with many everyday implications, the laws of classical physics do not apply and a curious world is revealed. 'Quanta' are the minimum amounts by which a property of a particle can change.

Particles, such as photons and electrons, are in some sense objects and in another sense waves. If you set out to measure them as objects, objects they will be, and if you set out to measure waves, that is what you will find. They are both objects and waves at the same time, the observation of one causing the disappearance of the other. When a particle moves from one place to another it is impossible to predict the precise path it will follow; in fact, it follows all the paths available to it simultaneously but emerges at the end of only one of them.

Particles can exist as pairs whose properties complement each other even if they become separated. In 1935 Einstein, Boris Podolsky and Nathan Rosen devised a thought experiment, now known as the 'EPR experiment', to try to discredit some of the uncomfortable implications of quantum theory. They asked what would happen if a particle consisting of two protons decayed, sending the protons in opposite directions. The properties of the protons would remain indeterminate until they were measured. In other words, each would be in a superposition of all its possible states and travelling in all possible directions. Because they form a complementary pair, however, measuring a particular property of one instantly determines the (complementary) property of the other. If one was found to be heading north, the other must be heading south. Since the properties were indeterminate until the instant they were observed, how does 'knowledge' of the state of the observed particle reach the other particle, regardless of the distance between them, which in principle might amount to billions of light years? Einstein and his colleagues maintained that communication is impossible, because it would need to travel faster than the speed of light. Some thirty years later they were proved wrong. The effect, called 'nonlocality', exists and has been observed experimentally.[27, 28] It is explained as 'holism', the word being used in a very specialized sense, that the paired particles form part of a whole system that remains connected at all times.

The most famous aspect of quantum theory is the uncertainty principle, discovered by Werner Heisenberg (1901–76), for which he received the Nobel Prize for Physics in 1932. This defines the limits of accuracy with which the position and momentum of a particle, such as an electron, can be measured. The uncertainty occurs because, to measure either characteristic, photons of light must be used and these impart energy to the particle being studied, which alters its behaviour. To measure position requires the use of light of a very short wavelength because the wavelength (the distance between one wave peak and the next) determines the distance within which the particle can be located. The wavelength of light is related to its frequency (the number of wave peaks passing a point each second) and the higher the frequency the more energy the photons have. Imparted to the particle, this energy alters its momentum. To measure momentum, light of a longer wavelength must be used to minimize the disturbance, but this restricts the accuracy with which position can be measured. In other words, momentum and position cannot be measured simultaneously with equal accuracy, because of limitations inherent in the act of measurement.[29]

Fascinating though quantum theory undoubtedly is, classical explanations for most everyday events remain perfectly adequate. It is hard to imagine how a system of beliefs, or new religious interpretation of the world, might be derived from it and parallels that have been drawn between quantum theory and the teachings of Asian religions seem at best superficial. Danah Zohar and Ian Marshall have adopted a different approach.[30] Using quantum theory as a metaphor for human society, they have derived ethical principles from quantum principles. They have not invented a new religion, but rather proposed that human behaviour should model itself on the behaviour observed at the most fundamental level of the real universe. The idea is interesting, if not altogether convincing.

More often, physicists have tried to explore the meaning of God in the light of physical laws. Paul Davies has written extensively on this topic.[31] Today it is much harder to do this than in past centuries, when it seemed obvious that all events were caused and causes preceded their effects. God could be explained as an original cause, the prime mover who set the universe in motion. This raises the question of how God came into being, of course, and if the answer is that the existence of God is necessary and uncaused, then the same argument might be extended to the universe, so that its existence is

27

necessary and uncaused, removing God from the picture altogether. Modern physics, and especially quantum theory, cast serious doubt on these types of argument, because its inherent uncertainties call into question our conventional ideas about causality. The combination of quantum and relativity theories produces a universe in which distinctions between matter and energy become fuzzy. For example, a collision between two particles moving at a certain speed can produce four particles moving more slowly, by the conversion of the energy of motion into matter. Nuclear reactors generate heat by the conversion of matter into energy.

To other people, things are simpler and Davies is needlessly mystical. There is no mystery about the origin of the laws of physics, according to John Sparkes: 'They come from the physicists who created them and they have the form they have because they provide the best basis so far devised for explaining and predicting the behaviour of that part of the universe which physicists study.'[32] Their limitation arises from the restricted domain to which physicists are confined. They do not extend to personal experiences or relationships or to human behaviour.

In fact, scientists *are* actively engaged in exploring the relationships between science, religion and morality. All the people mentioned here so far are scientists. Danah Zohar is a physicist. Most emphatically, it is not a matter that scientists consider unimportant or on which they are silent. Don Cupitt, the theologian, has joined the debate with a half-serious suggestion of an image of the universe comprising only flux. Energy and matter stream constantly from and return to an initial singularity so that the universe has no beginning, no end, and is continuously being destroyed and renewed.[33]

However, none of these people are attempting to contrive a new religion. Others have tried and have immediately run into two major problems. The first concerns mythology, the language in which the view of the universe is expressed. Those who would adapt Asian religions to western needs must overcome the profound differences in the cultures from which Asian and western religions arose. Although Buddhism and Hinduism have long attracted some westerners, they are still only a tiny minority. If an Asian religion is to be made comprehensible to Europeans and Americans of European ancestry, all of its tenets need to be restated in terms of European history and ideas. That would be a formidable task, which might yield nothing more than vague superficialities, the merest shadow of the original. The original is too deeply rooted in, and inseparable from, the culture that produced it.

An alternative, even more audacious approach, would be to invent a new religion entirely from scratch: principles, tenets, observances, everything. Edward Goldsmith has spent years doing precisely that and has outlined what he calls 'The Way,' though he defines religion differently to the way others do.[34] Explanation, or location of the individual, is irrelevant. 'Religion,' Goldsmith says, 'I shall take as the control-mechanism of a stable society.'[35] Its purpose, he claims, is to 'ensure that a society's basic structure is maintained'.[36] Consequently, it matters little what stories we make up by way of justification, so Goldsmith supplies his own, based on reverence for Gaia and the preservation of order within the biosphere as an overriding goal. The concoction of a religion solely for the purpose of social control must surely count as the most irreligious act imaginable. Genuinely religious people would be right to be appalled, were not the whole thing so ludicrous. In any case, it seems unlikely that many worshippers will be drawn to such shrines as may be erected to this fictitious but jealous god of the eco-fascists.

Goldsmith's is but one of the many contrived religions (see chapter 6) that specifically reject the description of the universe supplied by scientific discoveries. They can make no serious contribution to the quest for religious restatement, because sooner or later the need for explanation, for location, will arise and the explanations they offer will contradict other interpretations based on more profound thought that is supported by evidence and rational argument.

Meanwhile the serious quest continues. It is one in which scientists generally and physicists in particular still play the leading role that they assumed many years ago. But as was stressed earlier, the issues cannot be resolved by scientists alone, because they extend far beyond the boundaries of all scientific disciplines. The task is to provide a mythology that embraces what we know or are likely to discover about the universe and explains the place and role of each individual within it. Arguably it is a job for scientists, theologians and poets, working in concert.

3

The Quest for Certainty

No matter what mythology we develop to help us locate ourselves in the universe, serious difficulties remain. First, our concept of ourselves, what we mean when we speak in the first person, to whom it is that 'I' refers. 'As scientific knowledge progressed to colonize the entire universe,' writes Bryan Appleyard, 'the self became a safe refuge. In here we were safe from invasion and we found a home, an escape from the eternal wandering offered by science . . . We are consoled by the solitude of knowing we can never quite be explained, invaded or controlled by the world; we are terrified by the sense in which the world is indifferent to our fate.'[1] This is the crux of Appleyard's argument, that scientific information lacks significance: it tells us about all kinds of things, but remains silent on matters that are really important to us, matters relating to the way we feel, the way we perceive ourselves, about who we are. If the essential qualification for a scientific theory is its predictive power, then its contribution to our knowledge of ourselves in particular and of people in general is woefully inadequate. Predictions of the behaviour of systems that depend wholly on decisions made by people, such as stock markets, are no better than could be achieved by chance. The mathematical theory of chaos has thrown some light on such matters, but offers little immediate promise of improvement.[2]

It is far from certain that everyone is so self-absorbed as Appleyard suggests. Personally, I am interested in many things outside myself, including things that have no tangible effect on me, and I consider them quite important. I find deep introspection boring and unproductive. However, my own predilection apart, two points are being made here: we know very little about what constitutes the self; and our attempts to predict human behaviour are hopelessly unreliable. The second point is a common and sometimes dangerous fallacy that will be considered in chapter 12. The first point is the one that concerns us here.

Mind, body and self

The sharp distinction between self and non-self, mind and matter, subjective and objective, is usually traced back to the scepticism of the sixteenth century. Francis Bacon (1561–1626) was strongly opposed to theoretical speculation. The remedy he proposed was to anchor reason to experience. This is sometimes regarded – incorrectly – as the starting point for the modern scientific method, in which observation and experiment lead to the accumulation of facts. It does not work that way. Facts imply concepts, the intellectual framework into which the facts can be fitted, and it is this framework that guides the activities leading to the acquisition of facts. Were we simply to observe and accumulate the results of our observations, our work would lack any sense of direction, our facts would be incoherent. The philosopher Sir Karl Popper used to make this point by instructing students attending his lectures to take their notebooks and pencils, observe carefully and record their observations, so that at their next meeting all the observations could be discussed. Invariably, the bewildered students would ask: 'Observe what?' The observer must decide what to observe.

This was understood very clearly by Bacon's younger contemporary René Descartes (1596–1650). He thought the description of the world being compiled by scientists was confused because of the many individual contributions that had been made to it over very many years. The only way to introduce order was to scrap all of it and begin again and 'never to accept anything as true if I had not evident knowledge of its being so'.[3] This required him to reject all his ideas and opinions and to give free rein to pure reason. Noting that our senses sometimes deceive us, Descartes says, 'I decided to feign that everything that had entered my mind hitherto was no more true than the illusions of dreams. But immediately upon this I noticed that while I was trying to think everything false, it must needs be that I, who was thinking this, was something. And observing that this truth "I am thinking, therefore I exist" was so solid and secure that the most extravagant suppositions of the sceptics could not overthrow it, I judged that I need not scruple to accept it as the first principle of philosophy that I was seeking.'[4]

His first principle led Descartes to differentiate sharply between himself, the thinker of thoughts, and everything external to himself. This is the dualism for which he is renowned. One of its consequences was to represent everything other than the mind of the thinker as mechanical, the human body as an automaton under the direction of the mind. Mind and body, therefore, were each made of quite different stuff; they were qualitatively distinct.

31

Although today it is often criticized, the image of the universe as a machine composed of other machines proved immensely useful and remains influential still. Many of the concerns of environmentalists, for example, arise from studies of climate systems and ecosystems which treat them as machines.

While Cartesian dualism allowed free exploration of the universe, the mind remained mysterious. Only recently has it begun to yield up its secrets and, although it would be premature to suppose that the mind–body problem has been resolved, research has thrown light upon it that in time may lead to a resolution. The evidence so far will bring little comfort to the opponents of the scientific method. For 'mind' we can also read 'soul', 'consciousness' and 'sense of self'. The terms may have subtly different meanings for some people, but they are commonly used interchangeably. The first step in unravelling what now looks like a confusion introduced by Descartes is to abandon his dualism. The mind is not made of different stuff from the brain but is a feature of it, a physical organ. The brain is a machine, though the analogy between brains and machines or, more fashionably, computers fails to reflect the sheer complexity of the brain.[5] While we are used to thinking of the brain as an organ, many neurologists are now more inclined to regard it as a gland, a structure that secretes hormones and functions in response to them.[6] This further implies that the brain should not be thought of as confined to the skull, but as scattered throughout the body.

We are all aware of drugs that modify perception (hallucinogens, for example) or mood (such as tranquillizers and antidepressants). Work is now advanced on drugs that will go much further. Soon it will be possible to modify personality and character, to stimulate motivation, to enhance the feeling of self-esteem and generally to make people feel better about themselves and their lives. The efficacy of such drugs shows how far our concept of self is determined chemically. Obviously the feasibility of medication of this type raises important ethical questions that should be addressed before the drugs are introduced.

Thus, it is no longer a question of whether a machine could be conscious but rather of what kind of machine could be conscious. But what do we mean by 'conscious'? Clearly, consciousness must be a product of chemically mediated neuronal activity. The mathematician Roger Penrose has suggested this activity may involve quantum events, introducing a degree of indeterminacy into mental processes, which he uses to argue against the feasibility of our constructing intelligent machines.[7] Danah Zohar and Ian Marshall take his idea further, proposing a concept of consciousness that is derived from quantum theory.[8] There is widespread agreement, however, that consciousness, or mind, is a process, not a thing. Exploring that process, some years ago Francis Crick and Christof Koch began to investigate the visual system in an attempt to discover how mental events, such as the attribution of three dimensions to surfaces, can be caused by the firing of neurons; how 'blanks' in the perceptual image come to be filled in; for example, seeing a person from behind and inferring that that person has a face.[9]

There is a story among Buddhists of a young monk who visited a famous master who had been meditating in a cave for some years. With much difficulty, the monk eventually persuaded him to come out and begged him to pacify his mind.

'All right,' said the master, grudgingly. 'Give me your mind and I'll pacify it.'

'That's the problem,' said the young monk. 'I've searched everywhere and I can't find it.'

'There you are, then,' replied the master. 'It is pacified.'

The mind does not exist in the way we are used to thinking of it, as a distinct entity. That is what the philosopher Gilbert Ryle called a 'category mistake', leading to the idea of 'the ghost in the machine'.[10]

Another illustration may help. Imagine you have taken a guest to watch a game of baseball. Your guest has never seen the game played before and has no idea of its rules, so you set out to explain. You point out each of the players in turn and describe what that player aims to do. But at the end your guest asks: 'But where is the team spirit I've heard so much about?' The team spirit is not a thing, says Ryle, but 'the keenness with which each of the special tasks is performed, and performing a task keenly is not performing two tasks'. A player does not first catch a ball and then exhibit team spirit, or sometimes catch and sometimes exhibit team spirit. To divide catching and team spirit into separate activities is to make a category mistake. In the same way, mind, soul or consciousness is not separate from the brain but a product of what the brain does.

We are brought back to the question of how we define ourselves. What do I mean by 'me'? Clearly, my body is part of me but not the whole. I could lose limbs or certain organs, have other organs replaced by transplants, and my sense of identity would remain undiminished. I might be paralysed, so that I lost most bodily sensation and all independence of movement, but I would still think of myself as 'me'. In any case, my body has changed over the years. A person who last saw me when I was a small boy, 'me' as I was then, might not recognize 'me' as I am today. In the intervening years every cell in my body has been replaced several times. More likely, 'I' am my personal history, perhaps determined in part by small variations in brain chemistry that differentiate 'me' from all the other 'me's, but more importantly derived from contingency. 'I' happened to be born to these parents in this place at this time, for example. So I am my experiences and my memories of my own past experiences.[11]

The end of heaven, hell and the afterlife

The sense of self thus derives from memory, which is a function of the brain, and it is built and maintained by experiences, some delivered to the brain by our sense organs, others generated, as mental events, within the brain itself. When I die, my brain and sense organs will cease to operate and it is difficult to see how my sense of self can continue to exist. Once dead, I will have no functioning sense organs, so I will be unable to see, hear, touch, taste or smell anything at all: I will cease to be aware. Nor will I have thoughts or memories, since these will disintegrate as my brain disintegrates. I cannot survive my own death, for there is nothing left to sustain my self-identity. If the dualist concept of a mind separate from the body is false, then all talk of an afterlife becomes meaningless in a literal sense, for nothing remains that could survive, though the idea retains a mythical meaning.

At one time this thought would have been profoundly distressing to many people. Today it is probably less disturbing, because we have grown accustomed to it. When pressed, few of us really believe in the afterlife. Nevertheless, its loss has a certain moral significance. It was suggested in the last chapter that acts performed under the influence of promises of rewards in heaven or threats of damnation in hell were morally dubious. It now appears that the promises and threats are, in any case, hollow. If there is no afterlife, 'souls' cannot be consigned to 'heaven' or to 'hell'. Again, I doubt if this troubles many people, for literal belief in heaven and hell died long ago and, of course, it is perfectly possible to reinterpret these myths in terms of our immediate or future psychological well-being.

The demolition of the Cartesian mind–body dualism has been accomplished mainly by philosophers, with contributions from scientists, and it has been in progress at least since the early eighteenth century. Its final resolution will depend on a more detailed account of the brain than is available at present; this may allow scientists to explain fully the nature of consciousness. At the same time it will resolve another issue: that of consciousness in non-humans. Present evidence suggests that chimpanzees are conscious in the way humans are conscious, though perhaps not with the same degree of self-awareness, but we may hope to discover whether or not other animals are conscious. This will be of considerable ethical importance, for it will (or at any rate ought to) affect our attitude towards them and how we treat them.

The self is not the safe haven that Appleyard supposes. We cannot delimit enquiry and the brain is a legitimate subject for investigation. Even if all discussion of the nature of mind were forbidden, a definition of it would almost certainly emerge as a by-product of research towards a better understanding and treatment of brain disorders. We cannot delimit enquiry, if only because we will never agree on boundaries, but the very idea of doing so is wrong in principle. We have no choice but to come to terms with the implications of our scientific discoveries, however difficult or painful that may be. Neither can we pick and choose, accepting this bit of scientific evidence but not that bit, merely on the basis of personal preference. It comes as a package and, once absorbed, the difficult parts may be less disturbing than at first they seemed. Consider the number of impressionable people, especially children, who over the centuries have been terrified by descriptions of the tortures that they would suffer in hell if they failed to conform to patterns of behaviour invented and imposed on them by those in authority. There can be little doubt that the exposure of heaven and hell as mere fantasies is a benign and liberating product of rational thought, though some people believe it has led to moral degradation. By the same token, the abandonment of the idea of an afterlife is similarly liberating. It helps concentrate the mind on events in the real universe, which certainly are worthy of our attention, and imparts a greater sense of urgency. Thus, it frees us to contemplate progress towards a real and better future for ourselves and for those who follow us.

To the extent that anti-scientists persist in holding to beliefs that are no longer tenable, they isolate themselves from this enterprise. Far from contributing to a solution, they become part of the problem. Heaven and hell were once thought of as real places and were shown on maps. If we are to locate ourselves, literally or psychologically, it helps to have a reliable map, one that remains constant. A map that changes even as we study it, on which places spring into existence unpredictably, move around, and vanish without warning, is of little use to us. Historically, religion has supplied a stable map for our psychological use. True, it has changed over the years, but even geographical maps must be updated from time to time. Indeed, at one time the two types of map were closely related. Jerusalem was shown as the centre of the world, with heaven and hell above and below it. Even after the theological and geographical cartographers had gone their separate ways, religion continued to offer security. It could do so because the theological landscape was supposedly drawn with divine guidance, giving it an absolute authority that required no reference to anything outside the revelation on which it was based. Some people may not have liked it, but at least we all knew where we stood.

More recently, as close examination of the real universe made the religious map seem implausible, people have looked to scientists for a new certainty to replace the one taken from them. They might be forgiven for supposing that such certainty is possible. 'It's a scientific fact,' is often heard in bar-room discussions while, at the same time, advertisers try to woo us with the 'scientifically proven' efficacy of their products. Moreover the results of scientific studies are sometimes misreported to sensationalize them, so what is generally presented and accepted as fact is far from the truth.

Media distortions

In July 1993, newspapers reported that scientists had discovered a 'homo-sexual gene'. Suddenly there were anguished cries from politicians for legislation to protect individuals from discrimination, for example by employers or insurance companies, that might result from tendencies to homosexuality, alcoholism, mental disorder or any other ill that might one day be identifiable from a pinprick of blood. On 16 July, the day that the story broke, 30 British Members of Parliament signed an Early Day Motion drawn up by David Alton calling for a 'gene charter'. Within the gay community, some welcomed the news that their sexual orientation was biologically predetermined, while others feared it might lead to their oppression. Dr Nisson Shulman, British Chief Rabbi, supported the view expressed by some of his colleagues that should it become possible to genetically engineer the removal of the 'homosexuality gene', he would favour allowing Jews to volunteer for the treatment.[12]

At least one newspaper, The Daily Telegraph, had the good sense to invite a professional biologist, Richard Dawkins, to explain what had really hap-pened. The truth and its implications were far less dramatic than at first supposed. A team of scientists at the National Institutes of Health, in Bethesda, Maryland, led by Dean Hamer, had discovered that gay men were more likely to have homosexual brothers, homosexual maternal uncles and homosexual cousins on their mothers' side than would be predicted by chance.[13] The study involved 40 pairs of gay brothers, and mothers from 14 of their families, who had volunteered to take part. The result were purely statistical; the volunteers were not representative of all gay men and were not meant to be.[14] Near one end of the X chromosome, they found five genetic markers that were identical and shared by 33 of the 40 pairs of brothers. Although this suggested a genetic component to homosexuality, for which there was other supporting evidence, it did not amount to the discovery of a 'homosexuality gene'. Such a discovery could be claimed only after studies of more families and, as Dawkins carefully explained, it would have no social or moral implications whatever.[15] Certainly, genes map, precisely and deterministically, for the construction of proteins, but in the development of organs and whole individuals the rates at which genes are expressed, combined with environmental influences both inside and outside the body, produce an extremely complex pattern. The most that can be said about the genetic link to homosexuality, or any other behavioural trait for that matter, is that some people are inherently more likely than others to develop the behaviour. The possessor of a 'homosexuality gene' will not necessarily become homosexual, and a homosexual need not possess the gene. 'Whether you hate homosexuals or whether you love them,' concluded Dawkins, 'whether you want to lock them up or "cure" them, your reasons had better have nothing to do with genes. Rather admit to prejudiced emotion than speciously drag genes in where they do not belong.'

Had the study been of something other than homosexuality it might have been ignored. But coming as it did, at a time when controversy was raging over the admission of gays into the United States armed forces, it was bound to provoke a sensation. The story of the 'gay gene' is just one instance of external factors, often based on wishful thinking, seriously distorting the public perception of a scientific report. In this case, at least, no fortunes were to be made. However, the most tentative suggestion from a nutritionist, say of a dietary link to some illness, can be enough to trigger an avalanche of self-styled 'experts' peddling expensive nostrums, all of which have been 'scientifically proven', needless to say.

Scientists give the impression of constantly changing their minds and incessantly contradicting one another. Such confusion among the supposedly well informed provides ample material for cartoonists and comedians, but gives no comfort at all to those seeking certainty from their scientists, a map of the universe which, albeit incomplete, is at least stable in those areas it has so far managed to delineate. Those who look to scientists for the kind of certainty formerly supplied by religion look in vain. There are certainties, of course, at the most general level, for example, that the species of plants and animals we see about us have evolved from ancestors somewhat different from themselves; or that planets and satellites are maintained in their orbits by gravitational force. It is certain that genes encode for the construction of proteins: they can be seen doing it under controlled laboratory conditions. At the frontiers of research, however, certainty of this kind is not even a goal. To assume that research findings published in a scientific journal represent authoritative statements of absolute truth is to misunderstand the way that scientists work.

Scientists aim to explain the phenomena that they observe. Their reports, therefore, contain observations and attempts at an explanation. They are located within a context, defined partly by scientific and partly by social or political criteria. For example, there is currently intense interest in past climates, the movement of carbon between plants and animals, the atmosphere and the oceans, and in the transfer of heat by the oceans from the tropics to the poles. The subjects are interesting in their own right, but topical because of concern about the possibility of climate change induced by human activity, a possibility raised by scientists themselves. Topicality also influences funding, increasing the chances that a 'fashionable' project will be financed. Most scientists do not enjoy the luxury of working in intellectual isolation, devising the experiments and investigations that are most likely to satisfy their personal curiosity.

However, having embarked on a project, the scientist aims to be objective and impartial. The degree to which objectivity and impartiality can be realized is much greater than many anti-scientists suppose. The errors of experimenters who fail to take adequate precautions to exclude their own prejudices or irrelevant factors that might lead them to false results or false interpretations will usually be revealed. Either they will be noticed in their own descriptions of their research or their results will prove impossible to repeat.

Once a project, or a phase of one, is complete, the scientist writes a description of it in a rather formal, stylized document called a paper. There are clearly defined rules of presentation, which include proper attribution of previous work by others, so a three-page article usually ends with 20 to 30 references to earlier papers by other scientists. These are very important because counts are recorded of the number of times papers are cited. The distinctive style of scientific papers makes it impossible to publish them in newspapers or popular magazines; they must be interpreted for non-scientific readers and it is in the interpretation that most misunderstandings occur.

Scientists are under great pressure to publish, because their productivity is measured by the number of papers they write and promotion goes to the most productive. Most are deeply cautious however, because their papers are invariably subjected to the most rigorous scrutiny. This begins before publication. No reputable journal will publish a paper that has not been read and commented on by referees in the process called 'peer review'. Referees can question results and their interpretation, and can and often do ask for more information and for changes to be made. Only when they are satisfied will the paper be accepted, whereupon publication exposes it to the critical examination of other scientists working in the same field.

No scientific paper pretends to be definitive. It is an explanation of one, usually small, aspect of a phenomenon, or an observation, and its conclusions are quite likely to be revised as more information is obtained. The search is iterative and many papers are quickly superseded and forgotten. Very little is certain and, in principle, every scientific finding or statement may be overturned at some time in the future. Many have been, and the catalogue of abandoned explanations, consistent with what had been observed and at one time considered powerful, is a long one.

The case against certainty

Scientists prize honesty and intellectual rigour as the highest of virtues. It is not an offence for a scientist to be wrong. Mistakes can be made and can be admitted without fear of ridicule or opprobrium. Lack of rigour is sufficient to cause a paper to be rejected, but occasionally a researcher may falsify results in a way that escapes detection by referees. This is dishonest and so is plagiarism. These are the gravest offences that a scientist can commit and in serious cases are enough to warrant the dismissal of the offender from the scientific community. To err is human, but to cheat is unforgivable.

Not only is scientific information provisional, much of it is also in the form of statistics, and so is liable to misinterpretation. In particular, non-scientists who have difficulty with statistics can easily be misled. Statistically, a heavy cigarette smoker is more likely than a non-smoker to develop lung cancer. This is a probability, however, which is not disproved by the existence of heavy smokers who fail to develop cancer or the non-smokers who do, any more than a coin falling heads-up six times in succession disproves the likelihood of its falling heads-up for only half the number of tosses in a long run. Nor does the run of six heads alter the 50 per cent probability of the coin falling heads-up again on the seventh throw. Nevertheless, ignorance of the way probabilities work appears to offer a loophole for those who want to continue with their unwise habits or dismiss the results of scientific findings.

As has been stressed already, certainty is not to be found in scientific journals, only probabilities and explorations of ideas. But the absence of certainty is due not to some deficiency among scientists but is central to the scientific philosophy. Scientists agree on the existence of a reality exterior to the individual: a real universe that really exists. This reality, moreover, behaves in certain ways that can be described in general rules that are valid everywhere and are not violated, so in principle the reality can be explained in rational terms. On these matters there is no doubt. The uncertainty consists only in the provisional nature of scientific observations and explanations. To suppose that it brings into question the existence of reality itself or the feasibility of explaining it is to misinterpret it. Nor does the uncertainty support the relativist supposition that all explanations are equally valid. They are not, because a scientific explanation, unlike others, must be supported by the context from which it is drawn. It must form part of a larger, coherent totality. This is why many pseudo-scientific ideas can be immediately dismissed: they have no place in the coherent totality so far revealed. In many cases, the explanation must also be verifiable by experiment or through further observation.

The lack of certainty is also liberating. The history of this century surely supplies evidence enough of the harm caused by certainty and in previous centuries, when people were more certain, the situation was even worse. Certainty is irrational. Nazism was founded on irrationality and its central tenets were anti-scientific, but the Nazis were very certain of themselves. The members of the Branch Davidian cult were all well educated and of above average intelligence, but they were extremely sure of their own beliefs and, on 20 April 1993, most of them died for these ideas in Waco, Texas.[16] It is in the name of certainty that people persecute one another; if they doubted, if they left the smallest possibility of recognizing an alternative view, it is difficult to imagine how the persecution could proceed. Certainty closes minds and people with closed minds are dangerous.

Being human, some scientists may, of course, be susceptible to some kinds of certainty, but it is their professional uncertainty that allows their enquiries to proceed. This should not be a cause for fear, or of insecurity, but for the hope of continuing progress towards a more complete understanding of ourselves and the universe we inhabit.

II

THE FLIGHT FROM REASON

4

The Denial of Progress

The 'p' word is out of fashion. Utter it in polite society and you will appear naive and old-fashioned, even reactionary. Modern, sophisticated, enlightened people (those who, in former times, might have been called 'progressive', but let it pass) reject the very possibility of progress. Indeed, the word itself has largely fallen into disuse. It has no place in the conversation or thoughts of those to whom the idea that it is impossible is now so deeply ingrained as to be taken for granted. It is meaningless.

This is a fairly recent development. Once, horse-drawn vehicles were common and, as I recall as a boy, the dominant smells among the arches behind railway stations were of smoke and horses. Little by little the horse-drawn wagons were replaced by lorries. We called this 'progress', though for some years the highway was still referred to as the 'oss road'. People were happy with 'progress'. To them it amounted to improvement. Goods were moved more quickly and conveniently. When I was six I contracted mastoiditis, an extremely painful, severe and sometimes life-threatening ear infection. It was treated by surgery, which was also painful. Today the condition responds to antibiotics. The treatment is cheaper and safer and the victim suffers much less than I did. I am not embarrassed to call this 'progress' and I recognize that had I lived a couple of centuries ago, before surgery could be performed under general anaesthetic, my suffering would have been much greater and I might well have died either from the illness or from secondary infections arising from its treatment that were due to causes then unknown. That, too, is 'progress'.

We were quite clear about the meaning we attached to the word. Our concept of 'progress' was founded on the recognition that our lives were better than those of our parents and grandparents had been. On that basis, we anticipated that our children and grandchildren would be healthier, better fed, better housed and generally better off than we were. It was entirely reasonable that expectations should be raised. At that time, we were at first enduring and then emerging from the Second World War, and memories of the Depression were still fresh in our minds. We could see for ourselves that conditions were improving and were inspired by the view, held by most British people in the 1940s, that by working together we could build a new and better society. The past was described as 'the bad old days' by a generation of people who had been given cause to hope.

We were not much given to theorizing, otherwise we might have recognized that our optimism was based on the idea that history is a linear process. That is to say, we assumed each innovation was built on what had gone before, so our advance was essentially irreversible. Admittedly, there might well be setbacks, false trails that had to be abandoned, temporary losses of impetus, but over a timespan of generations history, we believed, would be seen to have a clear direction and past conditions could not recur: history did not repeat itself. We called the direction 'forwards' and the forward march of history was what we called 'progress'.

Can history be cyclical or chaotic?

The linear theory of history is not the only possible one. In *The Republic*, Plato distinguished between 'that which is sometimes generated and destroyed' and 'geometrical knowledge [which] is of that which always is'.[1] Things are created and destroyed according to the 'revolutions of the heavenly bodies'.[2] In other words, creation and destruction occur cyclically and, therefore, history itself is cyclical and success will attend the endeavours of those who adjust their activities to these cycles. Aristotle in his *Politics* criticized Plato, not for his cyclical theory of history, but for what Aristotle saw as inconsistencies in Plato's interpretation of it. 'Does a thing that came into existence on the day before the close of a cycle change simultaneously with things that originated long before?', Aristotle asked.[3] His own theory was certainly cyclical, for he described how one system of government would generate opposition to itself, leading to its replacement by an alternative, in a series that returned repeatedly to its starting place.

It is not difficult to see how history could come to be seen as cyclical. The seasons are repeated in an endless cycle, all the living organisms visible to the naked eye experience birth, growth, decline, and death – a linear experience for the individual that is rendered cyclical for the population or species by reproduction, and, of course, the rotation and orbit of the Earth produce a cyclically changing pattern of heavenly bodies. If we regard human societies as 'natural' in the same sense that these other phenomena are natural, they might well pursue cyclical paths of development and decline.

A cyclical theory of history is essentially static, in that such progress as may be achieved must necessarily lead to its own regress. Eventually, all that is gained must be lost. The theory may be unappealing, but that is insufficient ground for declaring it wrong, and there are cycles, other than those I have mentioned, which clearly do affect human affairs. For example, in parts of the United States droughts, it is believed, occur cyclically, and glacial and interglacial climatic episodes are thought to be triggered by cyclical changes in the amount of solar radiation received by the Earth, due to eccentricities in our planet's rotation and orbit. This theory, which is now widely accepted by scientists, was proposed in 1930 by the Serbian mathematician and physicist Milutin Milankovich (1879–1958) and bears his name.[4]

Cycles on the Milankovich scale may very well influence human affairs in the future, but they cannot have had any influence in history because the present (Flandrian) interglacial began about 10,000 years ago, before recorded history. Smaller climatic changes, such as the Little Ice Age, a cold spell that lasted from about 1550 to 1860, certainly affected people and possibly events but cannot be said to have altered the course of history. Meanwhile, our knowledge of past events, which vastly exceeds what was known to Plato and Aristotle, does not support the cyclical theory. Empires and even civilizations come and go, but no cyclical pattern is discernible. The same empire or civilization does not recover once it has fallen.

Alternatively, perhaps history has no pattern at all, but is chaotic. If so, events lead to other events, trends are established, but nothing endures, and outside influences randomly initiate or terminate sequences of developments. The mathematical theory of chaos, despite its possibly misleading popular name, deals with patterns that are definite but extremely complex, and in some sense may be almost cyclical. The two concepts are not so far from one another as it may seem, once the idea of very simple, obvious cycles is rejected. There is even a Platonic dimension to chaos, in that invisible forms seem to act as templates for patterns that can be observed.[5]

Which historical theory we choose is crucially important, for it conditions our attitude to the world and to our position and influence in it. Should we decide that history occurs in cycles, then our achievements become pointless, for inevitably they will be undone. Should we decide history is essentially random and chaotic, our inability to predict the outcome of our actions will inhibit what we do. We may even be afraid to act at all, lest the consequences prove harmful to ourselves or our descendants. Only the belief in history as a linear process can justify any optimism for the future.

Francis Fukuyama emphasizes this point strongly.[6] His thesis, 'the end of history', derives from the observation that a linear process may be expected to have an end, just as it had a beginning. However, his conclusion, that the end has been reached, is unconvincing and it is not even certain that the line need reach an end at all in the sense he means. True, the continuing increase of solar output will eventually render this planet uninhabitable, but even that will not bring history to an end if by then humans have settled elsewhere in the galaxy, which they might conceivably do. This colonization will necessitate the development of new social and political arrangements, so history will continue or will open a new chapter.

Is progress possible?

The linear theory and the concept of progress embedded within it are originally Christian. Some two thousand years ago, Christianity introduced two new ideas that were deeply subversive of the philosophical basis of the Roman culture dominant at the time. The first was that all human beings are equal in the sight of God, from which it follows that our perceptions of differences based on birth, gender, wealth or accomplishment have no moral significance, but are merely social conventions or political inventions that serve to ensure the retention of power by a ruling elite. The second was that history is linear, beginning with the Creation and ending with the Last Judgement. Christians further believe that we are all contaminated by the original sin of Adam and Eve. These ideas combine to produce what was once a radically new view of the individual, that each is born the moral equal of all others. Our lives proceed as a linear series of events; since history is not cyclical, each event is unique and irrevocable. We are born essentially sinful, but with the opportunity to conduct our lives so as to acquire the grace that will liberate us from that original sin: we can redeem and improve ourselves. In other words, we can progress and, by implication, so can the societies that we build. It was Christianity that introduced the concept of 'progress' to human thought.

This idea, described above as 'once radical,' is radical no more because it proved so powerful as to become central to the moral and political principles on which our institutions are based. Its power was greatly enhanced by a method for examining natural objects and phenomena that was introduced in ancient Greece, and is most closely associated in our minds with the names of Plato and Aristotle. The Greeks called it 'philosophy' (the love of wisdom). A century ago there were shops in London selling 'philosophical instruments', but nowadays this branch of study is usually called 'science' (literally 'knowledge', from the Latin *scire*, 'to know').

Scientists are potential revolutionaries. Hardly surprising then that they inspire fear among the timid and among the defendants of entrenched intellectual positions. Down the years, their correct interpretations of accumulated knowledge have overturned many traditional and deeply cherished beliefs. They have exposed earlier descriptions of the world and the universe as false. Perhaps it is the ongoing challenge presented by scientists that has contributed to the recent lack of trust in them by people unable or unwilling to grasp and accept new ideas, coupled with misunderstanding about the way that they work.

This distrust may account, in part, for the popular rejection of the idea of scientific 'progress', but more significant is the rejection of industrial or technological development. Scientists take much of the blame because of widespread confusion between science and technology. Today they are often closely related, as scientific discoveries can be rapidly followed by their commercial or industrial exploitation. However, this a recent development. For most of history, engineers have made devices with little or no understanding of the scientific basis of their inventions. Indeed, scientists advanced their knowledge by seeking explanations for the way those devices worked. 'Progress' has come to be associated with industry and its technologies. The rejection of the concept of progress is really a rejection of technological innovation. It has been blamed for the pollution of the environment and proliferation of most of the ills to which living organisms are susceptible. Rejectionists often express themselves in extreme language. Here is what Lewis Mumford wrote more than thirty years ago:

> Cybernetics, medical psychiatry, artificial insemination, surgery and chemotherapy have given the rulers of men the power to create obedient automatons, under remote control, with just enough mind left to replace the machine when its cost would be prohibitive . . . Another century of such 'progress' may work irreparable damage upon the human race.[7]

This is, of course, alarmist nonsense. It was nonsense then, but Mumford was not alone in thinking it. During the 1960s and 1970s whole libraries of books and countless newspaper and magazine articles were devoted to the so-called 'environmental crisis' facing our planet. Most of it was written in lurid language and dire consequences were predicted if the warnings were ignored. The underlying causes of the malaise that these writers claimed to have diagnosed were the expansion of industrial manufacture and technological innovation, or 'progress'. In 1970, Paul and Anne Ehrlich (not to be confused with the Paul Ehrlich of 'magic bullet' fame, see chapter 1) warned that 'Most of the great "advances" in technology from DDT and X-rays to automobiles and jet aircraft have caused serious problems for humanity.'[8] Interestingly, Lynn White, a historian, identified the origin of the 'crisis' with Christian belief. 'Our daily habits of action,' he wrote, 'are dominated by an implicit faith in perpetual progress which was unknown either to Greco-Roman antiquity or to the Orient. It is rooted in, and is indefensible apart from, Judaeo-Christian teleology.'[9] White's article offended Christians at the time because he linked Christians beliefs with the notion of progress, and the fashionable view of progress was that it was harmful.

Intellectual fashions change as arguments run out of steam. Journalists lose interest and readers turn their attention to other, more pressing matters. This particular campaign against progress ceased some time during the middle 1970s. Then, about a decade later, a new generation took up the cause, most of them quite unaware that the issues they 'exposed' were rather well known and their arguments had all been heard before. There were some differences, of course, in particular, that they had inherited the political clout of earlier campaigns. Governments and international institutions were primed to respond. Campaigners found it easier than their predecessors to challenge concepts of 'growth' and 'expansion', and to ensure, at least, the regulation of 'progress' if not its complete cessation, which the more extreme among them would have preferred.

Thus have we arrived at a curious position. The possibility of 'progress', on which our culture has long been based, is now denied. The denial accompanies a deep distrust of scientists along with, perhaps, a denial of the linear theory of history from which the idea of progress is partly derived and even of that aspect of the Christian tradition from which it arose. Championed by some journalists and politicians, these denials amount to a major revision in the way that many people view the world about them and their relationship to it.

To the true revolutionary, revolution must be perpetual. We might think those who have brought about this revolution in popular attitudes have achieved a remarkable success. But, of course, that is not enough to satisfy the zealot. We continue to be charged with backsliding into our antiquated dreams of social and material advancement based on a naive belief in what is scornfully dismissed as 'science' or, even worse, 'reductionist science'. Should we permit the revolution to consolidate its gains, far less to advance beyond them, we may find ourselves paying a price far higher than we imagine, with none of the benefits promised by the revolutionaries.

5

Decadence and Fear of the Future

Although the idea of progress is inherent in a western tradition that began some two thousand years ago, for much of that time it has been a fragile concept. Our history has passed through long periods when little changed and, during episodes when intellectual change did occur, usually the change was resisted. The establishment of progress, not merely as an abstract possibility, but as a natural and permanent feature, was based on the understanding of phenomena derived from the accumulated observations and interpretations of scientists. It was this understanding that gave people the confidence to innovate. Once that confidence was widespread, innovation of all kinds became possible. Economic, social and political traditions and institutions could be examined, and the assumptions behind them challenged. Schemes could be drawn up for reorganization based on more rational lines. Reform became feasible: in 1781 the word was used for the first time to mean a change for the better.[1] In a sense, the future was invented.

The Age of Enlightenment

The change began very slowly and for a long time it impinged little on the lives of ordinary people. Gradually, however, the accumulated understanding grew into a body of knowledge and ideas strong enough to challenge the prevailing theory of the universe. Just when that development occurred depends on the events that are considered the most significant, and is therefore a matter of opinion. However, there can be little doubt that a vast expansion in investigation, and acceleration in the rate at which new discoveries were made, began around the middle of the eighteenth century, in the period called the Enlightenment.

But still there were doubts and confusions. Around the 1740s, the idea of divine benevolence, that the world is as it is because God intended it so for the benefit of humans, proved attractive to many intellectuals across Europe, influencing individuals of most Christian persuasions. It bears a striking, if superficial, resemblance to the modern anthropic principle debated by cosmologists. Metals, it was suggested, are placed at distances from us according to our convenience: the most useful near the earth's surface, the less useful deeper down. The density of water, it claimed, is such as to permit ships to sail; the shape of the melon permits it to be cut easily into slices, and the teredo, or ship worm, performs the useful service of necessitating the continual repair and replacement of ships' timbers, without which felled timber would pile up uselessly.[2] (It is reminiscent of James Thurber's observation that 'wind is caused by trees waving their branches'.)

Such fanciful ideas faded away, leaving the philosophes free to argue and the Frenchman Denis Diderot free to try to compress all the technological information and political and religious opinions of the time into his Encyclopaedia. This is probably the most famous and, with sales of 4,000 copies, was perhaps the most successful book of that period. Not that everyone was impressed. Diderot was an atheist whose whole enterprise angered believers. Thomas Carlyle, for example, wrote in 1831 'that one day when the net-result of our European way of life comes to be summed up, this whole as yet so boundless concern of French Philosophism will dwindle into the thinnest of fractions, or vanish into nonentity!'[3]

Carlyle was mistaken, of course. The fame of the Encyclopaedia has endured, not least because, though there were many rivals and the idea was far from new, its success began a fashion for such works. For example, the title page of Encyclopaedia Britannica says that the first edition was published in 1768, 'by A Society of Gentlemen in Scotland', in whose honour to this day the page is decorated with a thistle. Carlyle was also wrong in underestimating the influence of the philosophes. The intellectual ferment released by their enquiries and debates gave rise to many publications, one of which, Common Sense, by Thomas Paine, the Norfolk-born editor of The Pennsylvania Magazine, was published on 10 January 1776. Within months it had persuaded all but the most conservative Americans of the need for political independence from Britain. Thus, it altered the course of the American Revolution, in which the first encounter had taken place at Lexington in November 1774. Even now, it stands as possibly the most forceful criticism ever written of monarchical government in general and that of Britain in particular. The American Revolution inspired the French Revolution, which began in 1789. Paine, who was in France in 1789, wrote The Rights of Man, dedicated to George

Washington but aimed at the British. Published in two parts, in 1791 and 1792, it was so successful and inspired so many people to dream of building a democracy in Britain, that for some years afterwards possession of a copy was taken by British courts as almost conclusive evidence of treason.[4]

In one respect, at least, Carlyle expressed a widely held view. He saw in the *Encyclopaedia* an attack on religion. Indeed, he called it 'the Acts and Epistles of the Parisian Church of Antichrist'. It was a similar accusation, in this case a most unjust one, that brought about the downfall of Thomas Paine. In *The Age of Reason*, published in two parts in 1794 and 1796, this deeply religious man sought to avert the risk he saw in France that, 'in the general wreck of superstition, of false systems of government, and false theology, we lose sight of morality, of humanity, and of the theology that is true.'[5] He had barely finished writing it before he was imprisoned by Robespierre for his criticism of the French and he narrowly escaped execution. His real target was not religion as such, so much as the clerical establishment and what he saw as over-reliance on the Bible. 'Whenever we read the obscene stories,' he wrote, 'the voluptuous debaucheries, the cruel and tortuous executions, the unrelenting vindictiveness, with which more than half the Bible is filled, it would be more consistent that we called it the word of a demon, than the word of God. It is a history of wickedness, that has served to corrupt and brutalize mankind; and, for my own part, I sincerely detest it, as I detest everything that is cruel.' For this display of apparent atheism he was vilified in both America and Britain. He died in poverty in New York City on 8 June 1809.

It is often said that more scientists have lived and worked during our own century than in the whole of past history. This is very probably true, and their number includes many intellectual giants. Yet the list of those active during the eighteenth century is impressive, with many familiar names: Isaac Newton (1642–1727), Gottfried Leibniz (1646–1716), Benjamin Franklin (1706–90), Immanuel Kant (1724–1804), James Watt (1736–1819), Jean Lamarck (1744–1829), Carl Linnaeus (1707–78) and the naturalist Gilbert White (1720–93). Others gave their names to units of measurement still in everyday use: Anders Celsius (1701–44), Gabriel Fahrenheit (1686–1736), André Ampère (1775–1836), Charles de Coulomb (1736–1806), and Alessandro Volta (1745–1827). Henry Cavendish (1731–1810) is remembered for the laboratory bearing his name, Edmond Halley (1656–1742) for the comet named after him, Luigi Galvani (1737–98) for the word 'galvanize' and Edward Jenner (1749–1823) for his introduction of vaccination. Leonhard Euler (1707–83) is said to have been the most prolific mathematician ever. Antony van Leeuwenhoek (1632–1723) was the first person to observe blood corpuscles, protozoa, bacteria, spermatozoa and many other microscopically small objects. James Hutton (1726–97) is often called 'the father of geology', and William Smith (1769–1839) discovered that geological strata could be dated by means of the fossils they contained, producing, in 1815, the first stratigraphic map of England and Wales. Others included the chemists Antoine Lavoisier (1743–94) and Joseph Priestley (1733–1804), physicists Jacques Charles (1746–1823) and Daniel Bernoulli (1700–82), mathematicians Pierre Laplace (1749–1827) and Jean Fourier (1768–1830) and astronomers Giovanni Cassini (1625–1712) and Frederick William Herschel (1738–1822). George Hadley (1685–1768) explained the atmospheric movements that produce the trade winds, Georges-Louis Buffon (1707–88) recognized that species of animals are not fixed but show variation and, in 1768, Joseph Banks (1743–1820) embarked for the South Pacific with James Cook (1728–79) on the first expedition ever to be organized and equipped for the collection of biological specimens.

These were the, so to speak, superstars of their age. But just as in the theatre and film worlds there are countless performers whose names are little known outside their own profession, so there were many lesser-known scientists of that period whose reputation lives on among present-day scientists, as well as many others engaged in the humdrum tasks involved in laboratory routine – the teachers and those researchers whose work came to nothing. The boundaries of scientific knowledge were expanding rapidly in many directions and the expansion continued and, of course, it continues still.

Probably, only very few, if any, of these people were atheists, and certainly many were devout Christians. Quite a lot of them were ordained ministers. So how did the challenge to religion arise? The conventional view is that progressively their findings combined to remove human beings from the position they had come to believe that they occupied at the centre of the universe. This demolition of human centrality, stone by stone and inch by inch, cast doubt on the centrality of humans in the mind of God and, since this was a cornerstone of religious belief, on the nature and, finally, existence of God. Obviously, there is some truth in all this, but probably the real threat was much more serious and this displacement was merely a manifestation of it.

Religion is based on a revealed truth, which is complete and divinely authorized. We may investigate the world about us, but only within bounds established by that divine authority, the results of our enquiries serving only to illuminate the work of God. As the scientific enterprise grew, so those divine bounds were forcibly expanded. It became increasingly evident that no natural phenomenon could be held immune from rigorous study and rational interpretation. God was not so much abolished as marginalized, for a long time into a 'God of the gaps' who retained dominion over only those phenomena for which scientists had not yet found an alternative, and more credible, explanation.

This, I think, explains part of the unease with which so many people confront scientific findings, attributing arrogance to those who dare to 'tinker' or 'meddle' with 'nature'. It is an attitude scientists find surprising and often incomprehensible. Joseph Campbell, not a scientist but the author of a major work on the mythologies of the world, wrote in 1970: 'One would never have thought, when I was a student back in the twenties, that in the seventies there would be intelligent people still wishing to hear and think about religion. We were all perfectly sure in those days that the world was through with religion. Science and reason were now in command'.[6]

It is only a partial explanation, however. If science were easy, its concepts and methods simple, there might be no problem in the replacement of one world-view by another. But it is not. Not only are the study and practice of science difficult (and, in our culture, often poorly taught), but the student must absorb vast amounts of information, accumulated over centuries, just to get started. The task is daunting and is not helped by condensing that information into a single volume that can be fitted into a curriculum crowded with other subjects. The intellectual overload leads not merely to the rejection of the study of scientific disciplines, but, in a culture to which those disciplines are so important, to a rejection of the whole culture. Later in the same essay Campbell says, 'in my own teaching I am today encountering more and more students who profess to find the whole history of our western culture "irrelevant". That is the brush-off term they use. The "kids" (as they like to call themselves) seem to lack the energy to encompass it all and press on.' That was in 1970. The trend he observed has continued, only now it affects not only students but a significant proportion of writers, journalists and commentators who have been educated in the humanities and seem to delight in denigrating scientists, though they know little about their work and understand less about the concepts underlying it.

Optimism vs. pessimism

The Christian tradition established the idea of history as a linear process, and hence of the possibility of progress. Christian doctrine permitted scientific investigation which undermined many of its tenets and substituted ideas based on the feasibility of improving society and the quality of our lives as well as our understanding of the universe. It is this that the detractors reject, the possibility of progress based on a rational interpretation of the circumstances in which we live. Since our 'invention' of the future was derived from the idea of progress, they also reject the future. 'One notes,' says Campbell, 'or at least sometimes suspects, a kind of failure of heart, a loss of nerve.'

Fear of the future engenders a deep pessimism. Many popular prognostications warn us that the future is certain to be much worse than the past, that through our intellectual (that is scientific) arrogance we are doomed. It should come as no surprise that this fear gives rise to preoccupation with the here and now, and the pursuit of pleasure and wealth for their own sake or as a source of emotional security.

A society that turns in on itself in this way is deeply troubled, for a general fear of the future is symptomatic of decadence. Admittedly, if we persist in this folly and permit the opponents of progress to consolidate their hold over us, the world will not come to an end. Scientific research will be curtailed, of course, because that is what they most strongly disapprove of. There will be fewer scientists and fewer people who are informed about the scientific view of the world around them, but life will go on. However, its quality for us will be gravely diminished. The effect will be local, confined to our own society, because belief in the future has been adopted throughout the world, and most people are not afraid of it. It is our own, fearful society, therefore, that will slip slowly and gently into the oblivion we have willed for it.

Increasing knowledge and improving ability to predict the consequences of our actions make it imperative for us to act responsibly. This imperative is fairly new. For most of history, people did what seemed to them would most likely deliver whatever immediate benefit they sought. While belief in the possibility of constructing the future encouraged optimism, that optimism was unconstrained. Common sense suggested that immediate benefits would endure for as long as they remained desirable.

The word 'optimist' first entered the English language in 1766. It derived from the doctrine of optimism propounded by Leibniz, who believed that any world is possible provided it is not illogical but that God considered all of these worlds before deciding to create the one we know. This world contains both good and evil, because some good is inextricably bound to evil. The relief of pain is good, for example, but that good cannot be realized without the prior existence of the pain, which is evil. A world that contains no evil is thus less good than a world in which there is the greatest possible excess of good over evil.[7] So God made the best of all possible worlds, a proposition that was widely interpreted to imply that this world was the one most conducive to immediate human happiness.[8] If that is what you believe, then you are an 'optimist' in the original sense of the word, the one satirized by Voltaire as the guiding philosophy of his character Dr Pangloss. This interpretation cannot be long sustained. As Bertrand Russell pointed out, such sanction from one of the greatest intellects of all time pleased the Queen of Prussia: 'Her serfs continued to suffer the evil, while she continued to enjoy the good.' Eventually, the word acquired its modern meaning of one who hopes for the best outcome, a sense in which it was first used in 1819.

The opposing view is most closely associated with the name of Arthur Schopenhauer (1788–1860). He maintained that the world and all it contains are manifestations of our will, which is evil, and the only good that the wise individual can achieve is through a kind of complete withdrawal and resignation. This gave 'pessimism' to the language, a word coined in 1794. Although his contribution is seldom acknowledged, the modern assumption that human activities almost invariably produce negative results owes much to Schopenhauer.

Rachel Carson, in her influential book *Silent Spring*, warned of such negative consequences arising from the careless use of modern pesticides.[9] The book was dedicated to Albert Schweitzer, and includes the quotation: 'Man has lost the capacity to foresee and to forestall. He will end by destroying the earth.' In 1963, Rachel Carson was awarded the Schweitzer Medal of the Animal Welfare Institute. The quotation is somewhat ironic, since the acquisition of detailed knowledge of our world allows us to make limited predictions, and *Silent Spring* was concerned wholly with predictions based on interpretation and extrapolation of scientific data (which Carson had selected carefully to support her thesis). It is ironic, too, because for most of his life Schweitzer (1875–1965) was an optimist and by no means averse to what I have called 'progress'. 'I acknowledge myself to be one who places all his confidence in rational thinking,' he wrote in 1931.[10] He welcomed the new road that passed through Gabon, where he lived and worked, built for strategic reasons in the 1940s as part of a highway to link Cape Town and Algiers. 'Of course these new roads are very imperfect and to travel on them is anything but a pleasure,' he said. 'But they signify a great step in advance over the old method of travel.'[11] Today it would be unthinkable for anyone dedicated, as Schweitzer was, to 'reverence for life' to applaud the building of any road, let alone one that spanned a continent and passed through the tropics.

The story of DDT

Rachel Carson warned of the dangers she foresaw of unrestricted and over-zealous use of insecticides. The possibility of abuse arose from the work of chemists, notably the Swiss, Paul Müller (1899–1965). His search for a compound that would be rapidly lethal to insects, but harmless to plants and warm-blooded animals, and that was sufficiently stable to stay in the environment long enough to minimize the quantity needed to control a pest infestation, led him to experiment with a substance first produced in 1873, by O. Zeidler. From it, Müller developed a compound called 1,1,1-trichloro-2,2-bis (dichlorodiphenyltrichloroethane). The compound was patented in 1943 and given the name DDT. It is worth recalling that DDT was introduced as a substitute for insecticides based on arsenic or nicotine, which could kill warm-blooded animals, and for vegetable compounds, such as derris and pyrethrum, which were obtainable only in limited amounts and rapidly lost their toxicity when exposed to air and sunlight.[12] For his discovery, in 1948 Müller was awarded the Nobel Prize for Physiology or Medicine (and gave away his prize money to help young scientists).

The first use of DDT was as a powder to kill the lice that transmit typhus to humans. Blown under the clothing of soldiers and millions of civilians, it prevented what would otherwise have been a severe outbreak of this disease. Some people suffered from slight itching, which might have been caused by louse bites, but there was no instance of serious illness. Later, there were a few cases of poisoning among people exposed to large amounts of DDT dissolved in oil or propanone (formerly known as acetone), but they all recovered. In fact, despite intensive research conducted over many years, DDT is not known to have ever caused the death or serious illness of any person. On the other hand, it is known to have saved millions of human lives. Rachel Carson, however, wrote: 'Dissolved in oil, as it usually is, DDT is definitely toxic . . . Once it has entered the body it is stored largely in organs rich in fatty substances . . . such as the adrenals, testes, or thyroid. Relatively large amounts are deposited in the liver, kidneys, and the fat of the large, protective mesenteries that enfold the intestines.'[13] At the time she wrote, it was not known that levels of DDT reach a maximum, after which the compound is excreted. It does not accumulate indefinitely and no long-term effects have been reported. The toxicity of DDT to humans is almost exactly similar to that of aspirin, but it is much more difficult to ingest a lethal dose of DDT. There were reports of DDT being found in human milk, but the highest level reported for whole milk (in Guatemala) was about 0.1 milligrams per litre; the figure actually reported, and the cause of considerable alarm, was that for milk fat, which was 30 times higher. DDT was also detected in the body fat of marine animals, such as seals, some as far away as Antarctica, and in the atmosphere. The concentrations were minute but knowledge of the presence of the insecticide caused widespread and quite unwarranted concern.

Nevertheless, there were problems with related compounds and adverse ecological effects from them and from DDT itself, arising mainly from their persistence and the fact that many insects, not only pest species, are susceptible to them. These effects had been known at least since 1945. Restrictions on the use of dieldrin and aldrin, the compounds with the most serious effects on wildlife, were introduced in Britain in 1961 and most uses of them and also of heptachlor were banned in 1965. In 1969 the Advisory Committee on Poisonous Substances Used in Agriculture and Food Storage recommended to the British Government that use of DDT be phased out in all but exceptional circumstances. The situation was far less dramatic than Rachel Carson suggested, at least in Britain, and those in possession of relevant scientific information cannot reasonably be accused of acting carelessly. The campaign against DDT in the United States continued for many years, with some writers claiming the public was being deliberately misled by 'toxic terrorists'.[14]

Eventually, the use of DDT was forbidden in most countries. The consequences of this are most fully documented from Sri Lanka, where spraying against *Anopheles culicifacies*, the mosquito that is the principal Asian malaria carrier, ceased in 1964. In 1948, when the spraying programme began, 2.8 million cases of malaria had been reported; in 1963 there were just 17 cases. Following the ban, by 1969 the number had risen again to 2.5 million. Thus the fear that by using DDT humans might be harmed and the environment damaged led quite directly to a great increase in harm to humans, in the form of malaria. There is no evidence that the ban on anti-malarial use brought any benefit to the Sri Lankan environment. When fear inhibits us from taking action, lest by that action we harm ourselves, our inaction may cause the injury we feared or even worse. In a sense, our prophecy of harm is made self-fulfilling. Interestingly, Dr Schweitzer welcomed the introduction of DDT, for the control of termites.[15]

DDT suffered from being the best known of a range of compounds, the organochlorines. It is also the least harmful. However, because of their persistence, all organochlorines can be accumulated along food chains until they reach harmful concentrations in predator species, and DDT is harmful to many aquatic organisms. The most celebrated incidence of such unintended harm arose from the spraying of Clear Lake, California, in 1949, 1954 and 1957, to control a small gnat (*Chaoborus astictopus*). Fish accumulated the insecticide and western grebes (*Aechmophorus occidentalis*) feeding on the lake fish were poisoned. Carson recounts the story and so does Bryan Appleyard in his *Understanding the Present*.[16]

The insecticide used was TDE (which Appleyard calls DDD), 'a close relation of the more notorious DDT', and he repeats the old warning of the dangers of TDE accumulating to harmful levels in human tissue. In fact, this never happened, though he is quoting from Carson, writing thirty years earlier. However, he does not mention something that she could not have known, that the poisoning incident yielded valuable information leading to modifications in pesticide use and improved conservation. The accident was unfortunate, but scientists learned from it. Far from regarding this as a benefit, Appleyard claims it as a 'fundamental challenge to the simplifying instincts of science', implying that humans should not intervene in the natural world at all, because its complexity 'defies the elementary mechanism we have tried to impose', in this case of a network of relationships among organisms that 'worked because of the operations of deep time' (whatever that means). Only later does he recount the use of DDT against malarial mosquitoes, suggesting an element of doubt over the wisdom of imposing so total a ban.

Towards a brighter outlook

Our attitude to DDT is but one of many examples of fear of the future. Another is the opposition to the controlled disposal of hazardous wastes, when the alternatives are more expensive and possibly more dangerous. The 'dumping' of low-level radioactive wastes in sealed containers in the deep ocean beyond the edges of continental shelves raises protests that the containers might leak, causing contamination in the food chain leading to fish species eaten by humans. However, the exchange of deep and surface oceanic water is very slow. In the Pacific, deep water remains isolated from surface waters for 1,000 to 1,600 years, and in the Atlantic and Indian Ocean for about half as long.[17] That is the time it would take for contaminants to enter the food chain leading to humans and it is long enough for the radioactivity to have decayed to insignificant levels. The alternative is disposal on land, as it is for all wastes formerly disposed of at sea.[18] As a means of preventing disposal in shallow or inshore waters, the ban is appropriate, but careful deep-water disposal is probably safer, and certainly cheaper, than any form of land disposal. Another example is the preference for sources of energy, such as wind or tidal power, that are presented as 'environmentally benign', but in truth are expensive and much more intrusive than those that their advocates fear.

Appleyard merely expresses a distrust of rationality that affects many people. Like most such commentators, he urges us to return to a system of beliefs that he supposes existed at some time in the past, when humans are presumed to have lived in some kind of intellectual innocence. We must not act, for the results of action are necessarily and invariably negative. We must not accumulate knowledge, for it will corrode our spirits. We must eschew logic, for it is arid and does not appeal to the emotions. With Schopenhauer, we must withdraw into a pessimistic quietude. 'Cutting down a tree in Brazil,' says Appleyard, 'is no longer an act of merely local significance, it threatens to kill us all,' though he does treat some apocalyptic warnings with more scepticism.

But whether we like it or not, we cannot withdraw from the world that we live in. We must breathe, we must eat, we must shelter ourselves and, in providing ourselves with the necessities of life, we act and our actions produce consequences. This being so, we produce the future as the consequences of our actions. That future is likely to be the one we anticipate or, perhaps more accurately, we are likely to select aspects of the future so that what we see tends to correspond to what we anticipate. Never mind that people live longer, healthier lives than did their parents; people are being poisoned and, paradoxically, too many babies are surviving. Never mind that our farms have never been more productive; the food is tainted and the farming destroys wildlife. Never mind that most of us drive cars, which require roads; roads cost us valuable farm land and destroy habitat for the wildlife associated with it. It is not the world that changes so much as our perception of it, and the current perception is profoundly pessimistic. We seem to find masochistic enjoyment in punishing ourselves.

Scientific research supplies us with reliable information on which to construe our understanding of the universe within which we act. The information may be insufficient for our needs, as in the case of the poisoning of Clear Lake, but this is hardly a criticism of scientists. The scientific project is not yet complete and by its nature is unlikely ever to be complete. Either we must act on the basis of the information available to us at the time or cease from all action, which is impossible. Nor should scientists be blamed because their findings have been abused. Those findings are published for all to read. Scientists did not invent human avarice or war. Evil did not originate with them and they do not claim to be able to abolish it. To blame scientists for failing to solve every problem facing us is not only unfair, it betrays a profound misunderstanding of the nature and purpose of scientific enquiry.

The unease of opponents of the scientific acquisition of knowledge seems to stem from the lack of reassurance they find in that information. To them, it presents a complex and confusing picture in which humans occupy no specially privileged position. From time to time, it may also reveal consequences of human actions that seem bad. Like the messengers of ancient Greece, scientists are castigated for this and their disciplines and methods, along with their findings, are dismissed as unhelpful or soulless. Again, such a view is founded in pessimism, to which there has always been an alternative. Armed with scientific understanding, we can improve our lives while minimizing any adverse side-effects. This implies that good outcomes are also feasible. It is the optimistic view, the view that allows us to believe in the possibility of progress towards a future we need not fear.

6

The Age of Aquarius

There is a sense in which Christianity introduced ideas that contained the seeds of destruction of some of its own tenets. By proposing the concept of linear time, which led to the possibility of knowledge accumulated through investigation and, consequently, progress, Christian belief made possible those discoveries that would lead to the destruction of its own mythology. This is not an anti-Christian statement, because Christian teaching has always encouraged rational enquiry. While faith is by definition irrational, Christianity opposes all other forms of irrationality, locating God outside space and time. This contrasts with those earlier religions that peopled the world with spiritual beings who sometimes appeared to chosen individuals in visions, but were otherwise invisible.

To deny the legitimacy of scientific enquiry and findings derived from it is not only to reject the optimistic ideas of progress towards a better future. It is also a rejection of the essential rationality of Christianity. In alliance, Christianity and scientific investigation are a powerful combination of a belief system and a factually informed base for comprehending the universe and our significance within it. Deny the alliance and either we dismiss the search for comprehension and location as unimportant or we adopt alternative sets of explanations. These usually involve magic.

Joseph Campbell once suggested that the visions of shamans, individuals who 'see' beings and events in a world invisible to others, so closely resembled the experiences of certain victims of schizophrenia that shamans might actually be schizophrenics, who lived in a culture that accorded value to their visions and so helped them come to terms with them.[1] This is an interesting suggestion, but one dating from the anti-psychiatry movement of the 1960s that would not be accepted by most psychiatrists today. Parallels remain, nevertheless, one of the diagnostic criteria for schizotypal personality disorders being 'magical thinking'. This is described as a belief in superstitions or in an ability for clairvoyance, telepathy or the possession of a developed 'sixth sense'.[2] Of course, this is only one of a number of criteria and such beliefs are not sufficient in themselves to warrant a diagnosis of a schizophrenia-type illness. What is interesting is that 'magical thinking' is regarded as pathological by scientists as well as by Christians.

I use the word 'magic' with a quite precise meaning. Bryan Appleyard writes of magic several times but in a different sense.[3] He considers it more or less synonymous with 'mystified wonder' or, perhaps, 'awe', in the face of the inexplicable, as in 'a magical sunset' or the 'magic' by which a tiny seed develops into a mighty tree. These are feelings with which most scientists are familiar and they are often intensified by the discovery of an explanation for a phenomenon. Contrary to the impression of science given by its opponents, each new discovery reveals new and greater wonders. 'Wonder', 'awe' and 'delight' are words that scientists often use.

Magic and the occult

I use 'magic' to mean a coherent system of belief according to which natural phenomena and forces outside the individual believer can be influenced by the performance of specified ritual acts. The corollary is that certain events not controlled by one individual may have resulted from interventions by another. Thus, someone may 'place a spell' on someone else or 'bewitch' their livestock or crops. Magic is manipulative and has little to do with passive appreciation. Such beliefs remain powerful in many parts of the world and they are emerging again in Europe and the United States after centuries during which Christian belief supplanted them.

The simplest demonstration of magical thinking can be found in the 'foolproof' gambling system. The game is one purely of chance. That is to say, the probability of a particular result can be calculated statistically. Play it enough times and, statistically, the result you seek will probably occur. That it will occur several times in one gaming session is improbable, but not statistically impossible. The systematic gambler may decide that the probability of winning is increased, say, by playing only on Thursdays. A win confirms the validity of the system. A loss does not disprove it, however. All it suggests is that the system is incomplete. The game should be played on a Thursday while wearing green socks. Again, a win validates the system, but a loss merely requires its further refinement. Perhaps an item of jewellery must be worn or not worn, or certain words must be spoken before playing. There is no end to the refinements that can be introduced, all the time elaborating an increasingly complex ritual that sometimes succeeds as, statistically, it is bound to do. Based on the premise that the outcome can be manipulated, the validity of the system cannot be disproved once that initial premise is accepted. The belief can be sustained indefinitely because at all times it remains detached from events outside its own system.

A person who rejects the fundamental scientific and Christian premise that natural phenomena can be comprehended rationally and, once comprehended, predicted, faces a world in which events occur arbitrarily according to the whim of supernatural forces ('supernatural' because they command natural events). Sometimes people benefit from those events, sometimes not, and so the supernatural forces have a direct relationship with human affairs. Floods, storms, fires and earthquakes are among the 'weapons' with which the 'spirits' or 'gods' punish human beings. Bountiful harvests reward them. Clearly, humans can influence events by behaving in certain ways. Propitiate the gods, avoid offending the spirits, and all will be well. Direct communication with gods and spirits being impossible, appropriate rituals must be devised and then refined. This is attempted manipulation of events and falls within my definition of 'magic'.

Go into any large bookshop in Britain today and you will find shelves devoted to works on magic and the occult. Often the shelf space given to these subjects far exceeds the space occupied by science books. In some towns (Glastonbury, in Somerset, is perhaps the most obvious example) there are bookshops selling only publications on magic, the occult and the quasi-religions that supply the systems of beliefs on which the rituals are based. You will even find books of spells for the use of 'witches', 'magicians' and the like as well as books on such esoteric topics as ley lines and divination. Alchemy is once again respectable, at least in these quarters, and Aleister Crowley is back in fashion. It was he who taught his acolytes the ever-convenient rule that 'do what thou wilt is the whole of the law'.

Reading such books is often heavy going. Since our language has developed in ways that allow us to describe the world around us it does not lend itself readily to descriptions of the fantastic. New words have been be coined or existing words given new meanings and, before long, the text degenerates into an impenetrable jargon. This accusation may seem unfair, since scientists experience similar difficulties and they, too, are often criticized for their reliance on jargon. There is a difference, however. Scientists use specialized terms, certainly, but only for clarity of meaning and brevity. The words and expressions used by them relate to the real universe and everyday language could be substituted. This must be so, because there are dictionaries in which specialist terms are explained in everyday language. It is more convenient to describe two species as 'sympatric', for example, than to say they occur in ranges that overlap, provided everyone agrees to use 'sympatric' consistently. It would be possible to conduct a scientific discussion without using technical terms, but the discussion would become very long, very tedious, and probably very muddled. A discussion of the occult, on the other hand, might be impossible if terms could not be invented as required to describe things which are remote from the tangible world.

Occultist jargon cannot be explained so simply, because it does not relate to common experience and definitions of jargon terms must themselves be written in jargon. The proliferation of terms feeds on itself. Given that the rise of occultism began several decades ago and authors are not content always to write simplified introductions to their topic, it is not surprising that the newcomer finds many of these books quite unreadable within their first few paragraphs. Any intelligent person can understand much of what they read in a scientific textbook, and even papers in scientific journals are often accessible to non-scientists who are prepared to be patient (though the editor of *Nature* has complained from time to time about the poor literary quality of papers submitted to journals).

The associated 'religions' are usually described as 'pre-Christian'. They refer to Druids, 'spiritual forces' of many kinds and the 'Earth Mother'. Witchcraft, involving divination and the casting of spells, appears to be thriving (I distinguish literal witchcraft, involving magic, from the use of the word 'witch' by feminist groups to draw attention to its former association with the oppression of women). There are groups that study teachings from 'Atlantis', and others whose leaders are guided telepathically, often by native Americans, now deceased. Despite their claims of antiquity, these beliefs bear only a loose connection with early religions. They have been compiled quite recently by people attempting to develop a universal theory without recourse to the rational thought they so distrust. One man founded an entire system on his profound observation that while walking he could see most of his own body but not his head.[4] The beliefs seem to be held sincerely. Many years ago I attended a conference at which an apparently sane and certainly sober middle-aged gentleman described, in a quiet, matter-of-fact way, his encounters with 'elementals' and meetings with the god Pan.

All of these developments are held to be signs of the dawning 'New Age' or the 'Age of Aquarius', linking them specifically to belief in astrology. Like other forms of divination, such as palmistry, Tarot card reading and the casting of runes, astrology is based on the idea that events are predetermined.

Although this is obviously paradoxical when combined with beliefs in the possibility of persuading emotionally unstable spirits in control of events to behave themselves, people have a keen desire to foretell the future. If events are predetermined, it is difficult to see what advantage could accrue to someone who knew what would happen but, by definition, could not alter or avoid it. Still, the desire exists and divination offers more. Indeed, its practitioners sometimes maintain that prediction is the least important aspect, arguing that it provides a valuable tool in explaining people to themselves and in helping them to reach decisions. The 'horoscopes' in newspapers and magazines, however, make this apology seem disingenuous. People read their 'stars' mainly for amusement but partly from a wish to know the future.

Retreat from the real world

Irrationality is widespread. Half of all the people participating in a British survey said they believed in faith healing and one-third of them consulted astrology charts. In the United States, one person in three of those surveyed claimed to have spoken with a deceased relative. One in five believed a rabbit's foot would bring them luck, and in Denver, Colorado, 150 people are counselled every day by telephone from the Mystic Quest Psychic Center. There are even university professors who believe in the power of pyramids.[5]

In Russia, too, there has been an upsurge of superstitious beliefs and cults based on them. Extrasensory perception is popular, as is the study of UFOs. In 1990, the then Soviet Academy of Sciences voted Alexander Spirkin to full membership. He had long been the principal philosopher to the Academy and was the author of a standard textbook. He was also a firm believer in ESP, clairvoyance, witch doctoring, and similar pseudosciences, which he urged should be studied in the same way as other phenomena and aspects of human experience.[6] Although they vary in details of their beliefs and rituals, adherents to these 'new-old' systems hold certain beliefs in common. They share, in particular, a desire 'to reclaim our links with the natural world . . . the world that . . . the pagans inhabited, and which many of us still do.' They talk of 'earth magic' and delight in being called 'pagans', believing that the original meaning of the word was 'country-dweller'.[8] In fact, it is derived from the Latin noun *paganus*, meaning 'civilian' as opposed to *miles*, 'soldier'. It acquired its present meaning of 'heathen' from the habit early Christians had of calling themselves 'soldiers (*milites*) of Christ', although it is true that a 'heathen' was once a 'dweller on the heath' and the adjective *paganus* meant 'rustic'.[9]

Some claim to be seeking a truth hidden beneath the world of phenomena, a meaning in nature to which they can relate, a natural law to which they can attune themselves. This truth can be apprehended only intuitively, an idea from which two conclusions can be drawn. First, that, far from elucidating the working of nature, rational thought and scientific investigation obscure it. Second, since some people are intuitively aware of the harmony of nature but others are not, those people are morally superior to others, although correct and sufficient performance of meditation and ritual may bring an awakening in the corrupted. To this extent, believers in the New or Aquarian Age are as prone to schism as any of their predecessors and just as given to exclusivity, defining themselves as belonging to a group of 'the saved' or 'the enlightened'. Excluded from the Christian in-group, the newly self-declared heathens have organized themselves into a group that excludes Christians and all rationalist outsiders.

Absorbed with the condition of their own souls and bodies, which they seek to purify by eating only foods uncontaminated by what they call 'chemicals',[10] the 'Aquarians' listen contentedly to their recordings of the natural music of streams babbling over stones, of the wind in the trees and the rustle of leaves, or of the mating calls of humpback whales. Their self-absorption is complete. Convinced that the world about them is inherently benign, that its ruling spirits smile on the affairs of those humans who feel naught but love for them, and that in great measure, direct action is the last thing they would contemplate. Superficially the most optimistic of religions, in truth theirs is founded on a pessimism as deep as that of Schopenhauer, because they hold that all human action results in harm to a world that could be made perfect if only humans would withdraw from it.

There is an alternative nihilism, of course. This is the one we see daily on our streets. It arises from the despair of people with no jobs, poorly housed or homeless, and with no prospect of improvement in their lot. Their pessimism is entirely rational and their frustration and desire for escape are easy to understand. Were their conditions to improve, however, hope would be restored to them. Their fear of the future is the product of their circumstances and does not imply a rejection of the concept of progress, only of their own exclusion from it.

At various times in the past, people have lost faith in what were the underpinning beliefs of their culture. Cast adrift, they have invented substitute beliefs, each more fantastic than the last and most looking to recreate a vanished past as an alternative to a future that promised to be worse than the present. They have sought refuge in ostentatious consumption, involving huge feasts and parties and flaunting costly possessions and decorations. They have demanded entertainments and, when they tired of them, have called for greater stimulus, with increasing emphasis on the infliction of physical pain or demonstration of sexual prowess. The societies manifesting such symptoms are usually called 'decadent'. Examples include Rome in the years following the death of Gaius Octavius (or Octavian), also known by the honorary name Augustus (63 BC–AD14), and France in the final decades of the nineteenth century. The novelist David Flusfeder has found parallels between those cultures and our own.[11] He takes the optimistic view that among the Aquarians, witches, magicians and others there may lurk the seeds of a new, vigorous revival. 'There is hope here,' he writes, 'because these are all attempts – no matter how silly – at transcendence, the cabbalistic urge to discover a greater truth, the belief that there is a greater truth to discover. In boredom comes transcendence. Barbarians today wear silly hats.'

It is more likely that such escapes from the real world and rational thought will lead nowhere. No big new idea has emerged from them to compare with that of hope for a better future. Nor have they devised a more effective means of comprehending the world than the one employed by scientists. Their 'love of nature' amounts to nothing more than sentimentalism, since it demands no effort on their part to acquaint themselves with living organisms and relationships among them, let alone the physics, chemistry and structure of the abiotic environment or the composition of matter. It is ignorance born of intellectual torpor. Despite its pompous pretensions to profundity it is devoid of content or meaning. It is also extremely dangerous. Since we must act in the world, sentimentalism or 'intuitive awareness' is a very inadequate guide. If we are to avoid the harm Aquarians fear, we must increase our understanding and reinstate our belief in the possibility of good outcomes. Scientists can help. Aquarians cannot.

If fear of the future drives New Age Aquarians into a deep introversion that allows them to inhabit a universe of their own making, it can cause others to retreat into the past, or into an idea of the past. This route attracts those whose commitment to one or other of the major religions prevents them from adopting an entirely new system of belief based on very different concepts, requiring not merely a denial of former beliefs but, by implication, a denunciation of them. Such people express their anti-rationalism by claiming the literal truth of sacred texts written long ago. They are usually known as 'fundamentalists', though so-called Islamic fundamentalism does not fall within this definition; that is based on a social and political interpretation of religion, rather than an approach to rational argument and scientific investigation. The term is often used pejoratively, those called 'fundamentalists' by outsiders preferring to regard themselves as 'traditionalists', in the sense of upholders of long-established values who affirm divinely revealed truths.

The triumph of creationism

For Christian fundamentalists the only relevant text is that contained in the Bible and, in particular, in the Old Testament. Their anti-scientific arguments centre on the conflict they see between the accounts of the creation given in the Book of Genesis and scientific findings.[12] Preferring the Genesis version, 'creationists' are driven to reject a biology that includes the evolution of species from ancestor species and the Earth sciences that assign ages to the Earth and its rocks far exceeding those that can be calculated from the Biblical text.

Although we are most familiar with Christian creationists, some orthodox Jews also insist on a literal interpretation of the Genesis stories. During the popular craze for dinosaurs that attended the film Jurassic Park in 1993, it was reported that a dairy company in Jerusalem had increased sales of its products by labelling them with a picture of a tyrannosaurus (though how this was connected with dairy produce was never clear!). Orthodox rabbis then threatened to withdraw the company's kosher certification because, according to Rabbi Zvi Gafner, 'dinosaurs are shown in encyclopaedias as beasts which lived millions of years ago, although it is known that the world was created 5,735 years ago . . . the dinosaurs symbolize a heresy of the Creation because they reflect Darwinistic theories.'[13]

In the United States, the proliferation of religious groups and the recognition in the First Amendment to the Constitution of the right of each citizen to freedom of religious belief and its expression, requires that schools funded by the Government may not teach any particular religion. Were they to do so, they would have to give equal weight to every religious opinion, which is clearly impossible. Creationists have sought to evade this restriction by inventing 'creation science'. This is not a science in the true sense, because it seeks only to discredit whatever contradicts its initial, fixed position. It concludes that 'honest seekers after truth must acknowledge that the evidence is overwhelming that man got here, not as a result of evolution, but by means of creation by God.'[14] The 'evidence', however, is selected, and often misinterpreted, simply to justify the Genesis account, which cannot be amended. There is no possibility of advancing the discussion and efforts to reconcile the two accounts are dismissed as 'false religion'.[15] If 'creation science' were genuinely scientific, its practitioners would rigorously test their alternative explanations and would welcome attempts from others to falsify them. They would also be prepared to revise those explanations in the light of informed criticisms. Since they do not do so, they cannot progress. This makes the case advanced by creationists and 'creation scientists' intellectually shallow. Indeed, the Biblical literalists exhibit a profound ignorance of the cultural context in which the books of the Bible were written and, surprisingly, of what the Bible actually says.[16] Nevertheless, a few examples demonstrate just how influential it has become.

Universities and other institutions of higher education are permitted to award degrees provided that their courses and the standards attained by degree candidates are acceptable to an authority maintained by the universities as a whole. This ensures that a degree from one university is academically equivalent to a degree awarded at the end of a similar course by any other institution within the network. In California, state law provides an alternative for institutions not accredited to the general academic system. Accordingly, the state education authorities are empowered to authorize the issuing of degrees by such institutions provided the course work and facilities they offer are comparable to those of accredited institutions.

In 1988, teams from the state education authorities paid the first of a series of investigative visits to one such institution, the Institute for Creation Research Graduate School (ICRGS), which until then had been issuing master's degrees in astrogeophysics, biology, geology and science education. The investigators reported to Bill Honig, the Superintendent of Public Instruction, recommending that support be withdrawn because the course work on offer fell far below acceptable academic standards.[17] The recommendation was accepted. Then, in January 1992, the decision was reversed. The state education authority agreed to pay the ICRGS $225,000 and to permit it to continue issuing master's degrees. Other similar institutions were also allowed to continue awarding degrees and diplomas. State lawyers said these institutions 'may continue to teach the creation model as being correct, provided the institution also teaches evolution'. Bill Honig was prevented from participating in further decisions concerning licences to the ICRGS.[18] At the ICRGS all subjects are taught from a creationist view point. Geology, for example, is taught by a person who believes the Earth to be less than 20,000 years old, all sedimentary strata to have been laid down during the Biblical Flood, and that Noah's Ark contained the animals from which those we see today are descended.

This was the most recent battle in a war being fought in several countries, but most publicly in the United States. It ended in victory for the creationists, who are especially powerful in California, where they are bidding seriously to take control of the education system. The war has been simmering away for more than a century and the final outcome remains uncertain. Having suffered reversals earlier in this century, culminating in the famous trial of John Scopes in Tennessee in 1925, in the 1980s creationists began to fight back in the United States using the same First Amendment that restricted them to justify the inclusion of particular minority views in school curricula. In December 1981, a further trial, described as 'Scopes Trial II', took place in Little Rock, Arkansas. Scopes had been tried for teaching that humans had descended from a lower order of animals. He was convicted, though the conviction was later overturned. Despite the popular impression that the case resulted in a victory for evolutionists, in the longer term it encouraged the resurgence of Biblical literalism.

'Scopes II' concerned the requirement, by Arkansas law, that equal classroom time be allocated to the teaching of biology and 'creation science'. It turned, in part, on whether or not all the strata of sedimentary rocks were deposited during a single event, the Biblical Flood. That battle was won by the scientists when the judge ruled that the law was unconstitutional because it forced biology teachers to teach religion in science classes.[19] It is not only what happens inside classrooms that is affected. For many years, American publishers, fearing their books might prove unacceptable to the state authorities responsible for central purchasing, demanded that school textbooks contained no mention of evolution or, at best, a brief and guarded outline of it. The battle is now being fought in Australia as well as the United States.[20]

Every culture has produced its own myths to account for the existence of the Earth and everything in it. People have always sought explanations for what they see about them. If they are sensible, and most people are, later understanding will cause them to abandon those explanations or to retain them as stories of historical, literary or allegorical value. These merit study, and their loss would be tragic, but this does not mean they should be accepted for all time as the literal truth even when they contradict the well-attested results of formal investigations using methods not available to their authors. They are 'traditional', of course, in the sense that they belong by right within

71

a cultural history, and someone who works to preserve them for that reason might justly be described as a 'traditionalist'. Scholars in Greece, Italy, Egypt and most other countries who seek to prevent stories from their own cultural histories from being lost may claim to be guarding traditions. But it would be absurd to expect them to believe in the literal truth of the stories they are preserving. That would amount to denying the culture itself, for cultures must live and develop. Without their dynamism they die, leaving only their fossilized traces. The true 'traditionalist' belongs to the mainstream of the culture, rejoicing in new ideas, the advancement of thought and the growth of understanding.

Most Christians are sensible people and, although discussion of evolution caused difficulties in the last century, they have long since been resolved. The Church of England never adopted any 'official line' on the matter and it was not necessary for the Protestant churches to do so either. They emphasized the personal relationship between the individual and religious truths, allowing each member to reach his or her own decision.[21] The more conservative and authoritarian Roman Catholic church, on the other hand, had to decide one way or the other and, as is its way, it took rather a long time and a series of papal encyclicals to make up its mind. In 1893 Pope Leo XIII took up a literalist position in *Providentissimus Dei*. Pope Pius XII moderated this in 1943, with *Divino Afflante*, which advised those interpreting the Old Testament to bear in mind the ways of thought of its authors. Finally, in 1951, the same pope moved just a little further, in *Humani Generis*, allowing Catholics to discuss evolution. The consequence of this was that Catholic teaching was freed from the risk of being proved wrong in matters susceptible to investigation, such as the age of the Earth and the evolution of species, while the belief that all humans are descended from an initial pair, which is much more difficult to prove or disprove, could be retained.[22] (However, progress is being made in that direction, too, through comparisons of sequences of mitochondrial DNA.)

Creationism and 'creation science' offer escapist creeds to bring comfort to the fearful and credulous. Like the beliefs of the New Age Aquarians, they represent a flight from reason, an alternative to serious thought. Believers must deny the evidence of their own eyes, for, despite claims to the contrary, new species are evolving from ancestral species at the present time. The most famous of many examples are the 'ring species'. The herring gull (*Larus argentatus*) is a common British bird. It also occurs in North America, but there it is slightly different. Continue westwards and by the time you reach Siberia some of the herring gulls are beginning to resemble the lesser black-backed gull. This resemblance increases, until by the time you return to Britain the lesser black-backed is recognized as a species, *Larus fuscus*, separate and clearly different from the herring gull, with which it does not interbreed,[23] although throughout most of the 'ring' the local gulls can and do breed with groups adjacent to them on either side. Ring species provide unambiguous evidence of the alteration of one species into another and emphatically disprove the creationist contention that the identity of each species is fixed and immutable.

The lure of Oriental religions

Yet another option is available to the seekers-out of alternative religions. They may embrace one of the Asian religions and many have, Buddhism and Hinduism being the most popular. These offer complete, coherent and rational systems of belief, but ones that differ radically from Christian belief. Their imagery is derived from natural and cultural environments very different from those familiar to Europeans or North Americans of European extraction and some adherents may be attracted by the appeal of the exotic. There is a much more profound difference, however. In both these religions history is regarded as cyclical. All events are governed by cycles of varying periods. Living beings die, their individuality disappears and, as a result of their deaths, other beings are born inheriting the accumulated consequences, or 'karma', of the beliefs, thoughts and deeds of their predecessors. Such progress as is possible consists only in the extirpation of this inheritance, leading to the complete annihilation of the concept of 'self' and allowing what is no longer a separate individual to merge seamlessly with a kind of universal unconsciousness.

It is a system of thought that simply denies the future and, with it, anything that could be recognized as progress. People are born to occupy a particular position within a rigid social order and cannot be considered as unique individuals. Education is based on the relationship between the student and a 'guru', the role-model that the student must emulate, absorbing instruction unquestioningly. To dispute the smallest particular of what is taught would be considered an act of rebellion against the cosmic order. As Joseph Campbell describes it, 'the principles of ego, free thought, free will, and self-responsible action are in those societies abhorred and rejected as antithetical to all that is natural, good, and true'.[24] Uncomfortably, perhaps, for those brought up in the individualistic west, if taken seriously these religions impose strict rules for everyday behaviour. Probably few people actually observe them, but old books of the laws stipulate precisely what must be done, and how it must be done, for every moment of every day, down to such details as the correct direction in which to face while sneezing or yawning. 'Every detail of life is prescribed to an iota, and there is so much that one *has* to do that there is no chance at all to pause and ask, "What would I like to do?"'[25]

Some people may find comfort in acceptance of such a discipline. It relieves them of all responsibility, save that of obedience to the rules. More especially, it removes from them the burden of independent thought and judgement. The world outside is illusory. Not only *may* it be ignored, it *must* be ignored, for its investigation can lead only to spiritually damaging illusion. Reality lies beyond and behind mere appearances. We may find it for ourselves, inside our heads, and that is the only place where we should seek it. Happily, the rules forbid the seekers after truth from studying critiques of philosophies based on solipsism.

In Asia itself, serious people have been working hard for a very long time to reach an accommodation between traditional systems of belief that deny the possibility of progress and the western belief in and hope for the future. They have had considerable success and much important scientific research is now conducted throughout that continent. This adds to the global store of knowledge that enriches everyone. Most of all, it enriches those societies in which it is produced.

Just as literalist beliefs derived from Christianity and Judaism are archaic, so too are the versions of Asian religions adopted in the west by people seeking an escape from the future. All of them are anti-rational, because the understanding they seek is not to be attained through observation and reasoned thought. They are also deeply pessimistic, because they deny the possibility of progress and improvement. The popularity of such archaisms, like that of the New Age Aquarian pseudo-religions, may also be regarded as a manifestation of decadence.

7

Monetarism and Idealism

The word materialism causes much confusion. To many people, it means the accumulation of wealth or material goods as an end in itself, as when we talk of 'living in a materialistic age' by way of comment on the proliferation of household gadgets and package holidays. This makes it more or less synonymous with 'greed'. Scientists are believed to contribute to this 'materialism' in two ways. First, since most people fail to distinguish between scientific research and technological development, scientists are held responsible for the production and marketing of the goods people seek to acquire. Second, because all scientific endeavour is guided by a philosophy founded on materialism, the feeding of avarice is seen as its logical consequence.

As understood by scientists, materialism has a quite different meaning, wholly unconnected with greed. It means, simply, that all natural phenomena may be explained in terms of the interactions of material substances under particular conditions. In many cases those conditions are represented by energy: gravitational attraction, the electromagnetic force involved in chemical reactions, temperature, pressure and so forth. At one level, matter and energy are interchangeable, because of the relativistic equivalence of energy and mass, but in all cases they are linked. The light that powers photosynthesis is produced by matter in the Sun, the pressure that deforms rocks is caused by gravitational attraction and the movements of other rocks, the heat of volcanoes is produced mainly by the decay of radioactive elements in the Earth's mantle. Assuming that there is a material explanation for phenomena is a precondition for scientific interpretations, and is not meant to extend beyond their province. If non-material explanations were allowed, they would not explain anything. The question would merely be transferred. If, say, 'spirits' cause something to happen, we are bound to ask what these spirits are, what they are made of, and how they exert their power. Unless answers can be found to those questions, enquiry becomes pointless. This materialism has nothing whatever to do with greed.

Nor does the materialism underlying scientific investigation necessarily imply the denial of spirituality of which scientists stand accused.[1] Certainly, scientists are trained to be sceptical and to demand that assertions purporting to be scientific be supported by evidence and logical argument. But there are many scientists who consider themselves religious and believe in a god, and only very few who are immune to a sense of wonder at the beauty that their work reveals.

Scientists are neither more nor less likely than anyone else to regard wealth as a good in itself and they cannot justly be accused of advancing such a belief. The fact remains, however, that our society does seem to base itself on that belief. We do consume more than our fair share of the world's resources and we are encouraged to attribute moral worth to the acquisition of wealth. We are set examples on which we seem to be expected to model our lives, of possessing many material goods and consuming at a level that our grand-parents would certainly have considered ostentatious, if not downright vulgar.

The change occurred fairly recently, forced by economic and political developments that had nothing to do with scientific research. We have established, or permitted to be established, an economic system that relies for its health on the production of consumer goods and hence on our consumption of them. It was in the late 1950s that the so-called 'consumer society' really came into existence. In the preceding years, western economies had been concerned mainly with the expansion of heavy industry, first to supply the needs arising from the Second World War, which began with industry suffering severely from the effects of the 1930s depression, and then to rebuild civilian manufacturing. Many people express concern about the ethics of 'consumerism', and they are justified in doing so. Apart from its wastefulness, it is based on manipulation. Vast advertising and merchandis-ing empires devote great skill and considerable expense to persuading us that we need goods that we do not need, before assuring us that the goods are manufactured only because we have demanded them. The familiar slogan 'the consumer is in charge' is very far from the truth.

Consumerism and liberal democracy

Our unease is warranted. Indeed, the change to consumerism may be seen as a turning away from the ideals derived from the Christian and scientific concepts of progress. They entailed spiritual, intellectual and social progress, which were regarded as being partly dependent on but inherently more valuable than material progress. Depending on your point of view, our new condition may be described as 'arrival' or as a temporary reactionary phase. Both views, and most commentators, associate this new condition with the spread of liberal democracy. This is the political philosophy most of us accept as the source of benign, tolerant institutions and governments that historically have proved stronger and more durable than those produced by the rival philosophies of fascism and communism. Communism is said to have failed, you will recall, because it proved unable to engender the growth of a consumerist society.

The liberalism in the phrase 'liberal democracy' is meant in its economic sense, of course, and does not necessarily imply tolerance of individuals or ideas that conflict with it. Nor is it notably benign in its treatment of those who fail to benefit as they should from the opportunities it is supposed to offer. A 'democracy' is defined as a country in which the government is elected by and answerable to its citizens. But, by this straightforward definition, not all liberal democracies are equally democratic. Spain under the Franco regime, Portugal under Salazar and apartheid South Africa were all liberal in the economic sense, but they were assuredly not democracies. Some states suffer from the corruption of politicians or the entire political system. The people of some-what democratic Britain are, strictly speaking, subjects not citizens. They have no written constitution setting out their rights and no opportunity to influence the choice of head of state, an hereditary office, or membership of the upper house of the legislature. This is filled partly by heredity, partly by appointment, and in the case of some Church of England bishops ex officio. As is usually the case, the accepted wisdom is a sweeping generalization.

Francis Fukuyama sees liberal democracy as the end point of our linear history, the destination towards which progress has been leading us.[2] While acknowledging the dangers of lapsing into complacent self-absorption on the one hand or of becoming a race of violently and bloodily embattled individualists on the other, Fukuyama welcomes the prospect of our arrival. It leads to an homogenization that will unite us all, as 'the apparent differences between peoples' "language of good and evil" will appear to be an artifact of their particular stage of historical development.'[3] Bryan Appleyard, in contrast, sees liberal democracy as something close to the ultimate evil, but defines liberalism as 'the form of society in which the government's only real concern is the maintenance of order, plurality and tolerance.'[4] His objection arises from the odd belief that liberalism denies the liberal person convictions of any kind.

J.K. Galbraith, in his own study of the development of liberal democracy or, if you prefer, the political and social ideas that arise in advanced market economies, provides a more convincing explanation: 'individuals and communities that are favoured in their economic, social and political condition attribute social virtue and political durability to that which they themselves enjoy'.[5] Allow the market to identify demands, create them where they do not exist and then satisfy those demands and a point will be reached where most people will enjoy a higher standard of living, and more material goods, than previous generations. They will be content with the economic and political system that satisfies them, says Galbraith, and will vote for its perpetuation. Not everyone in the community will benefit, of course, but the poor, unemployed, sick and otherwise disadvantaged will comprise a minority, effectively disenfranchised in elections. This sounds very much like what happened in Britain in four successive elections starting in 1979, and is not unlike the description by Fukuyama, who writes of the 'victory of the VCR'. However, Galbraith does not consider this situation permanent. The effort to satisfy immediately every whim of the consumer leads to contradictions that are ignored by the contented member of society until the system is challenged by economic disaster, defeat in an international war or revolt by the neglected underclass.

An anti-intellectual culture

A society in which a majority of people feel contented, or one that has reached the culmination of its historical development, may well lack any sense of purpose. Progress has come to an end and the future will stretch ahead as an interminable continuation of the present. There will be no future in any meaningful sense and fear of the future will, to some degree, be a fear of mind-numbing boredom. Those commentators who discern in this a cause of moral decline are correct, correct but hardly original. As long ago as the 1950s Richard M. Titmuss was warning anyone who would listen of the ways in which the removal of all constraints on the operation of the market economy was eroding moral concepts based on the strong caring for the weak.[6] The social consequences he predicted are now evident in the United States and Britain and in other European countries.

Whether we are rushing headlong toward contentedness for some or the end of history for us all, there is little reason to question the broad diagnosis. We are ruled by the politics of self-interest and acquisitiveness. Like everyone else in our society, scientists are not immune from the effects of this loss of moral purpose. Indeed, they are compelled to conform to its dictates, for it is from society at large that they obtain the funds for their research and to pay their salaries.

It was reported in August 1993, that the assessment of 'science linkage' showed British and American companies to be closer to the 'cutting edge of science' in their research than their Japanese counterparts.[7] This finding was based on the number of references contained in patent applications to papers in scientific journals or other publications, but the report also pointed out that a high number of citations does not guarantee commercial success. Significantly, the fruits of scientific research are judged less by the interest or intellectual value of the new knowledge they contribute than by the extent to which that knowledge can be translated quickly into monetary profit.

Scientists are now expected to acquire at least some commercial expertise. It has been proposed in Britain, for example, that the method for training research scientists to doctorate level be changed. Instead of spending three years as graduate students or research assistants serving a kind of apprenticeship in a laboratory, they should spend four years, the first of which would be devoted to acquiring a master's degree, which is partly taught. Most scientists welcomed the proposal in principle, as it would help identify potential researchers, while awarding a valuable higher degree to those who complete the first year successfully but are not suited for a career in research. Then it emerged that part of the master's degree would relate to a course in management. A knowledge of management and the relationship between scientific research and commercial advantage are to become requirements for scientists. So far as I know, no similar requirement has been proposed for musicians, actors or painters, though their work can also be evaluated in monetary terms.

Scientists cannot escape being drawn into the commercialist culture. Perhaps they should feel flattered that their work is seen as valuable, even by those for whom money is the only measure of worth. Perhaps some of them do. The majority of British scientists, however, are deeply demoralized, sensing a perversion of what they regard as their true purpose, the acquisition of knowledge and understanding for its own sake. There is a distinct political antipathy towards 'curiosity-led' research, which many scientists consider the most important part of their work.

This antipathy spreads to the public at large, where it lends support to a more general anti-intellectualism. George Walden, a former British minister responsible for government spending on civil research, described in the Daily

Telegraph how he was armed with the final chapter in Appleyard's book, cal 'The Humbling of Science', when he met physicists at CERN, the Europe centre for nuclear research. There, he said, he met 'soft-spoken zealots' w had 'a quasi-mystical belief in what they were doing'. Had he spoken in t way about the motives and work of novelists, poets, painters or musicians would rightly have been dismissed as a philistine. That some people fail recognize scientific creativity and dedication as arising from the same sou as artistic creativity and dedication should, by now, come as no surpri What is alarming is that such a person should have power over the lives a work of scientists.

Scientists as scapegoats

Popular distrust of intellectual activity is made worse in the case of scienti by shrill protest against almost any application of their findings. As c commentator on Walden's article pointed out, 'having been blamed . nuclear weapons, radioactivity in general, other environmental problems a the ethical dilemmas thrown up by the new biology, they [scientists] are n being blamed for the complexity and uncertainty of the modern worlc Scientists cannot win.

Clearly, problems exist. Millions of people are without employment. Th are children who are inadequately fed and clothed. Thousands of people sle in the open for want of accommodation of any kind. Once banished by a soci that believed in the possibility of progress and sought to achieve it, beggars ha returned to the streets of all our cities. People are attacked and robbed so t many are fearful of going outdoors after dark. The incidence of most categor of crime is increasing. In part, at least, these are the consequences of econon and social policies pursued by governments elected and re-elected by democratic process. In a world as complex and uncertain as the mode world is, there is a temptation to deny that any of this arises from decisions have made, is of our doing, and instead to seek scapegoats.

Scientists make convenient scapegoats, but to blame them is to indulge self-deceit. They are people, like us, members of the same society. As Hila and Steven Rose have demonstrated, 'much of our modern science has be shaped by the requirements and constraints placed upon it by the society which it is performed'. It does not 'fall upon society like a stone, mouldir bending, or crushing it.'[9] We get the scientists we deserve in the same se that we get the politicians we deserve.

Scapegoats represent what we fear in ourselves. Scientists stand for the ideal of knowledge as a source of beauty and understanding. They are rejected because we cannot appreciate that beauty and because we fear knowledge may bring power and the responsibility that goes with it. Scientists stand for progress, but progress seems impossible. They stand for the future, but the future seems frightening. Rejecting scientists, their values and their ways of thought makes us turn against rationalism in favour of the irrational. For some, truth is apprehended intuitively, and reality exists only inside their own heads. Having isolated themselves in solipsism, they seek to assert their individuality by showing off their wealth, believing 'if you've got it, flaunt it', and to hell with all those less fortunate.

This is the flight from reason into the darkness of deep pessimism. The rejection of the scientific ethos, of Christian rationality, of belief in the possibility of progress, leads to a dread of what tomorrow may bring. That, combined with the ostentatious consumption with which people seek reassurance for their existence, is a sign of social decline.

III

FANTASIES OF A JUVENILE APE

8

The Juvenile Ape

Humans are insatiably curious and also playful. These are traits we share with many mammals. Bring a cat into your home for the first time and once it begins to feel secure it will start to explore, eventually investigating every corner, indoors and out, until it has mapped its new surroundings completely. Turn a horse into a field it has never entered before and it will first explore the boundaries of the field and then every inch of the interior. Young horses play, mainly chasing games and their own version of 'follow-my-leader'; kittens and puppies play at pursuing and catching prey and many retain their playfulness into adulthood. Any new object or animal turning up inside the area they regard as their home range will be investigated carefully but cautiously. Among monkeys, apes and humans, the so-called 'higher' primates, exploration and playfulness assume a much more important role.

Can we draw legitimate parallels between our own behaviour and the behaviour we observe in non-humans? Many years ago I was a professional actor, like most actors spending years in repertory companies where we performed a new play each week. Once I had memorized my part I used to rehearse it repeatedly in my head. Mentally, I recited the lines to myself, made the appropriate moves, gestures and facial expressions, imagined myself simulating the emotions I was required to project. Each morning, when we rehearsed, all of us spoke the lines we had learned, moved and simulated a part of the performance in preparation for our appearance before an audience. This sounds obvious, for it is well known how plays are prepared and how actors work. But in fact we all do this. Everyone rehearses mentally, or by acting out in private, a task that promises to be difficult.

Now compare this. Jane Goodall, the scientist who, from the 1960s, devoted years to the study of chimpanzee behaviour, reported three separate occasions at Gombe, Tanzania, where she worked, when adolescent males performed charging displays (by which males seek to dominate subordinates) deep in the forest, well out of sight of their companions. She watched a four-year-old female watching her mother fishing for termites, then mimicking the action with a small twig. Goodall asks, how can we be sure that the displaying adolescents were not charging an imaginary crowd of chimpanzees, or that the infant, Wunda, was not catching imaginary ants in an imaginary ant nest?[1] Young children love to act out the scenes in their heads, with objects and persons only they can 'see'. Do chimpanzees do the same thing?

Like humans, chimpanzees are very curious about one another. An individual who does something new, something its companions have not seen before, attracts close attention and others may try the new behaviour for themselves. They learn from watching one another, so the entire group may acquire new behaviours which are then passed on to their offspring. This is the beginning of cultural evolution, the process by which humans transmit information from generation to generation.[2] There are two species of chimpanzees, Pan troglodytes, the 'common' chimpanzee, which is the species Jane Goodall studied, and the bonobo, Pan paniscus. Some of the scientists who study communication in chimpanzees believe that, of the two species, it is the bonobo (also known, misleadingly, as the 'pygmy' chimpanzee) that resembles humans more closely.

Chimpanzees are very different from humans, of course, but when Europeans first encountered them it was their similarity to humans that they found striking, not the differences. The first specimens were regarded as a 'puny race of mankind'. The name 'chimpanzee' was first used in 1738, taken, it was said, from an Angolan word meaning 'mockman'.[3] The orang utan (Pongo pygmaeus) is more arboreal than the African apes and essentially solitary, but it, too, has been thought of as a kind of human; the name 'orang utan' is Malay for 'man of the woods'. The name of the third genus of great apes, Gorilla, is Greek and first occurs in an account of a voyage of Hanno, a Carthaginian who explored the west coast of Africa in the fifth century BC. The name may derive from an African word meaning 'wild man'.

The evolution controversy

For many people, it was a misunderstanding of the proposed relationship between humans and non-human primates that aroused most hostility to the Darwinian theory of evolution by natural selection. Indeed, the issue continues to trouble some people. Not long ago I overheard a young man in a restaurant say to one of his companions: 'Evolution, who believes in that? We're descended from monkeys? I don't believe that, do you?' He appeared to be perfectly serious.

The controversy arose with the publication of *The Origin of Species* (1859), though, in fact, Darwin says nothing directly about human evolution in the book. It came to a head in the famous debate at the 1860 meeting of the British Association for the Advancement of Science, in Oxford, when the Bishop of Oxford, Samuel Wilberforce (1805–73), is alleged to have asked Thomas Henry Huxley (1825–95) whether it was 'through his grandfather or his grandmother that he claimed descent from a monkey'. At least, that is the story; whether the incident actually took place at all is uncertain.[4] Certainly, it was Huxley himself who first explored human evolutionary relationships in detail.[5] But neither Huxley nor any other serious biologist has ever suggested that 'humans are descended from monkeys'.

Biologists have two principal methods for comparing species. The first examines anatomical characteristics; the second compares genes using the techniques of molecular biology. Both methods seek to count the traits shared by different species and to distinguish between those they have all acquired from a common ancestor, called plesiomorphic, and those acquired more recently which distinguish the species from their ancestors. These are called apomorphic. The ancestor of the giraffe, for example, had a short neck. This is the plesiomorphic condition and the long neck of the modern giraffe is apomorphic. A third category, of synapomorphic, or shared-derived traits, describes apomorphic characteristics that are shared by two or more groups; the possession of such traits strongly suggests the groups are related. Molecular biologists use a similar system, but examine either sequences of genes or the composition of proteins. Since genes encode for proteins, identical proteins commonly imply identical genes. Synapomorphic traits are identified by comparing the groups being studied with an unrelated out-group. Comparisons of synapomorphies then allows taxonomists to construct a branching diagram, called a cladogram, showing the links between groups and their ancestors.

Species are grouped taxonomically into genera, genera into families, and families into orders, with sub-, infra-, and super- categories introduced as necessary between them. On this basis, extant members of the order Primates which are most closely related to humans comprise the suborder Haplorhini (the Old World groups), and within that the superfamily Hominoidea, which includes two families, the Hylobatidae (gibbons), and Hominidae. There are two subfamilies of Hominidae, the Ponginae, with the single genus *Pongo* (orang utan), and the Homininae, with three genera, *Gorilla*, *Pan* (chimpanzees) and *Homo* (humans). Colin Groves, of the Australian National University, has counted seven synapomorphies between gorillas and chimpanzees, 12 between gorillas and humans, and 25 between chimpanzees and humans.[6] Protein (DNA) studies confirm the general relationships, though there are differences over the dates at which evolutionary branches diverged.

Broadly, the current view is that the evolutionary line leading to the modern Old World species divided first, roughly 20 million years ago, with the separation of the line leading to the gibbons. About 15 million years ago it divided again, with a branch leading to the orang utan. A third branch, leading to the gorillas, separated about nine million years ago, and the chimpanzee and human lines separated about seven million years ago.[7] The branching that separates humans from chimpanzees occurred so recently in evolutionary terms and the genetic relationship between the two is so close that some scientists have suggested humans and chimpanzees should be regarded as sibling species and included in the same genus. Genetically, the difference between one human being and another is greater than that between a genetically average human being and an average chimpanzee.

The search for ancestors continues and, as new fossils are discovered, dated and classified and further molecular evidence emerges, more pieces are added to the jigsaw that is our own history. In years to come, no doubt adjustments will be made and dates refined or altered, but the general outline will remain. It shows that humans, chimpanzees and gorillas are all descended from a common ancestral species. It does not show, and no one has ever suggested, that we are in some way descended from chimpanzees or gorillas. Look at a chimpanzee, gorilla and human, however, and the chimpanzee and gorilla resemble one another much more closely than either of them resembles the human. The genetic distance between the species may be small, but clearly it is highly significant, or at least it appears to be. There may be an explanation for such a large effect arising from so small a cause.

Over the course of evolution it can happen that the rates at which an animal develops, grows and matures become dissociated. When this happens, a juvenile may resemble the adult of an ancestor species, or an adult may resemble an ancestral juvenile. Such an alteration in the rate of development is called heterochrony. The bonobo, for example, is about the same size as the common chimpanzee, it matures at the same age, goes bald when it is old like the common chimpanzee though perhaps at a slightly greater age, and it is born with the black face that the common chimpanzee develops only when it matures. The shape of its skull and the proportions of the parts of its body, however, resemble those of a juvenile common chimpanzee. This is a form of inhibited or long-delayed development called neoteny, and is one reason why humans find bonobos more attractive than common chimpanzees: their flattened faces are more like a human face.

Humans, then, share this foreshortening of the face, one effect of which is to bring the top of the nose to a position between the eyes. This is a neotenous feature and there are others. Our third molar teeth are reduced in size, for example, and there are several more technical features in the shape of the skull. Some biologists believe our lack of body hair and layer of subcutaneous fat are also neotenous features. Our body hair is much like the downy covering ('lanugo') of other infant primates. Our lips, too, remain adapted for sucking throughout life, unlike those of other primates. It may also be, though this is controversial, that neotenous changes in skull shape are associated with increasing cranial capacity.

The acquisition of neotenous features has not been a steady progression. Some species have been more neotenous than other, later ones and it is the overall trend that has been towards neoteny, rather than a steadily increasing neoteny over the course of our evolution as a whole. In *Homo sapiens neanderthalensis* (Neanderthal man), for example, growth continued for much longer before sexual maturity was attained, a condition known as hypermorphosis, leading to an adult possessing entirely novel features, but our own species, *Homo sapiens*, is the most markedly neotenous of all *Homo* species.[8]

The human ape

In a word, we are apes, but, in a sense, juvenile ones who never grow up entirely. This has certain consequences, the most important of which are psychological: we retain certain juvenile characteristics throughout our lives. Soon after birth, both human babies and baby chimpanzees show signs of recognizing objects, but a difference soon emerges. The chimpanzee is interested primarily in acquiring the object and examining it, but develops no particular interest in communicating anything about the object to anyone else; it seems uninterested in communication as such, though, of course, chimpanzees communicate fluently about things that affect them directly. The human infant, on the other hand, soon starts communicating to a parent the fact that he or she has noticed the object; the baby shows a desire to communicate.[9] That said, there is strong evidence that the psychological differences between humans and chimpanzees are much more a matter of degree than quality, and that degree centres on the acquisition of language. Whether or not chimpanzees can acquire language is a matter for debate, but it seems likely that they can do so to a limited extent. It is in humans, however, that language is most fully developed.

Jane Goodall describes how she was talking to a friend about chimpanzees and apples: 'By a strange coincidence she, as a little girl, had firmly believed in a fairy who lived in the oldest apple tree in her parents' orchard. Is this a real difference between man and chimpanzee, the *imagination* that creates fairies and Eve in the Garden of Eden?'[10] Perhaps that is part of the difference. The rest concerns language. Whereas chimpanzees show every sign of being able to form conceptual images, humans do so, but also frame them in language and then use them in stories – entirely fictitious constructions. Moreover, humans continue to play this game all their lives. It is a childish game, but it is our childishness that makes us human.

As juvenile apes with a talent for language, we love to play with stories. They are important to us and we use them to interpret the world about us, a world we never cease to examine and explore. From these stories we construct myths, which become absorbed in our culture so deeply that they come to resemble 'given truths', observations and explanations handed down from one generation to the next. We take them for granted, often quite unaware that the world we think we perceive in fact is described for us by our myths.

Myths are extremely important. They tell us things about ourselves, explaining human relationships that are commonplace but difficult and may cause suffering. They help us locate ourselves in society and the universe. Sometimes, though, it is necessary to abandon them because they no longer conform to what we know of the real universe. A myth that has lost its relevance little by little loses its credibility. Most of us, for example, no longer believe in the story of Adam and Eve, at least in a literal sense, or in the literal truth of classical stories about the intervention in human affairs of gods and demigods. Our abandonment of these myths does not mean we have abandoned all myth, however, only that those to which we hold now are buried more deeply in our psyche. Some of these myths are also irrelevant to the real world and it is time we challenged them, as we have challenged earlier ones, for it may be harmful to persist in believing a myth that describes unreality.

9

Golden Age and Noble Savage

At first, the world was a happy place. Only recently created, Epimetheus and his brother Prometheus had furnished its animals with wings, claws, shells and the other attributes appropriate to their needs. Epimetheus had made men, too, but by the time he got to them he had used up his stock of faculties and there were none left with which to endow them. That is why Prometheus, helped by Minerva, went secretly to Olympus, lit a torch at the fire of the Sun and brought the fire to men as the gift that would help them survive. Enraged at the theft, Zeus made Pandora, the first woman, and sent her to be the wife of Epimetheus. Ever inquisitive, she opened the jar in which Epimetheus kept all the plagues for which he had found no use. Epimetheus resealed the jar too slowly to prevent all of them, except hope, from escaping, but it was a long time before their effects were felt.

The completed world was filled with delights. There were no laws or judges, but justice invariably prevailed. Trees had not been felled and the forests were just as they had been created. There were towns, but they were unfortified, there being no weapons to assault them with. No one worked, because the land produced an abundance of everything anyone could need. It was always spring. Flowers bloomed by themselves, with no need for seeds, rivers flowed with milk and wine, honey seeped from oak trees. It was the Golden Age. Eventually it passed, of course, and we have been yearning for its return ever since. It is, perhaps, the most enduring of all our myths, as strong now as it ever was.

The paradise myth

The story occurs in most cultures and in surprisingly similar form. The Garden of Eden is one example. The name Eden means 'delight' and our word 'paradise' is derived from the Persian for 'a walled enclosure,' or garden. Eden, of course, was fed by four rivers, was well supplied with gold and onyx and produced food in abundance. Suffering comes later, when Eve and Adam had eaten of the fruit of the tree of knowledge of good and evil, and Pandora had opened the jar and released everything except hope. Obviously, these events never occurred in history, although the Eden story was believed literally until quite recently, and some people still believe it. If so widespread a tale is not an historical record, it must have another significance. It must tell us something not about the outside world but about ourselves. Its meaning must be psychological.[1]

Most of us are practical folk, dealing with a familiar world of everyday miseries, hardships and bad weather. Psychological meanings are too subtle for us, so the Golden Age has assumed a different meaning rooted in nostalgia. This, too, is psychological and apparently supported by evidence. I remember clearly, when I was a small boy the summers were much longer and sunnier than they are today, and the snow lay deep every winter. Most of us have similar memories, so they must be true and it is no wonder we believe the climate is deteriorating. The real truth, of course, is that we remember the exceptional and forget the ordinary. If you doubt this here is a little test. Without looking, describe in detail the design and wording on a coin – a 20-pence coin, for example, or a quarter. Most people fail the test, though we handle these coins every day. They are so ordinary that we take no notice of them.

The Golden Age thus becomes a past time when 'things were better'. Traditional morality prevailed, making the streets safe to walk in and allowing us to leave our doors unlocked without fear of intruders. We were more prosperous then and happier. The world was a better place. Whenever you hear someone advocating a reform that is intended to resurrect former conditions you can be sure they are in the grip of an irrational Golden-Age nostalgia. The nostrum is doomed to failure for two reasons. First, there never was a Golden Age. Second, different believers place the Golden Age in widely differing historical periods, so there is not the slightest hope of agreement.

For some, the Golden Age occurred in the latter part of the last century, when industry thrived and the idea of Social Darwinism was taking root (see chapter 12).[2] Others locate it in a much more remote time. Many eighteenth century writers believed it was the period when Rome was ruled by Antoninus Pius and his immediate descendants, in the second century AD. John Locke places it in a time when people lived in a 'state of nature', when they were guided by reason and so had no need for laws.[4] Plato thought it was in the days before the Mediterranean lands were deforested.[5] Kirkpatrick Sale suggests it may have been in the Athens of Pericles, despite its slavery and the subjugation of women: 'Yet for all its ills, that city at that time had such an appreciation of the central role of the human within the society, of individual worth mixed with communal value, of civic participation and reward, that a contemporary American could not observe it without some sense of what has been lost in the intervening years.'[6]

Although we usually think of the Golden Age as being in the past, utopian movements locate it in the future. Despairing of the past, they hope to build a new society, a new world, the like of which has not previously existed. The dream may prove incapable of realization, but it has the merit of being founded on hope for the future rather than nostalgia for an imagined past, and of believing firmly in the feasibility of progress towards its desired goal.

Utopianism is denied those who fear the future, where they see only dystopias, Today, when fear of the future is endemic, many people look backwards in their search for the ideal society. Since the quest begins from a tarnished present, it must take us back to the society that existed before those developments began that brought us to the present state of affairs.

In 1974 there was an explosion at a chemical plant at Flixborough, in Yorkshire, England. Analysing its causes, Edward Goldsmith concluded: 'the important and hence relevant "cause" is not, in fact, the building of the plant, but the creation of a situation in which it can be predicted that the plant will be built, i.e. the development of an industrial society'[7]. Once embarked on this route it is difficult to know where to stop, because it can be argued that our present condition is the direct consequence of our natural curiosity and inventiveness. To be consistent, therefore, we should locate the Golden Age at some time before the evolution of primates. In which case, no human being could have enjoyed it, and there is always a risk that some other species, perhaps of social dinosaurs, might have evolved intelligence and, from that, an industrial society. Caught in a regression of his own devising, Goldsmith chooses a place to stop. His choice is arbitrary but popular among those of a somewhat sentimental disposition. 'Tribal societies,' he says, 'are remarkably stable.'[8] He goes on to prescribe what members of his ideal society should believe: 'The society's protection is assured by the family gods and those of the clan and tribe – ancestral figures who . . . are . . . considered to have simply graduated to a more prestigious age-grade.'[9] In other words, we should consent to be governed by the elderly and the dead.

The noble savage

It is a most popular place, this world of the tribal society, to which many would journey. You will meet them at open-air festivals and deeply green conventions, and will recognize them by their uniform of supposedly 'ethnic' garments. When they arrive at their promised land, the travellers will join the existing inhabitants, belonging to a race of mythical beings, the noble savages. Meanwhile, the savages must be protected by all means from the contamination of our industrial society. We know of them mainly from Jean-Jacques Rousseau (1712–78), one of the Enlightenment philosophers:

> Among the savages personal interest speaks as strongly as among us, but it does not say the same things. The love of society and the care for common protection are the only bonds which unite them. The word property, which causes so many crimes to our honest people, has almost no meaning among them. They do not have any discussion of interests which divide them. Nothing leads them to deceive one another. Public esteem is the only good to which they aspire and which they value.[10]

The noble savage is of ancient lineage. Probably the line began in Greek tradition, with dog-headed bogeymen who lived in India – not the country as we know it today, but a vague, distant land inhabited by monsters. The tradition reaches us through a rendering of classical texts by the Greek patriarch Photius (c. 820–91). He described them as a mountain people who 'do no work; they live by hunting, and when they have killed their game, they bake it in the sun . . . These Kynokephaloi have no houses, but live in caves . . . They wear a light dress of trimmed hides, as thin as possible, men and women alike . . .'[11] By the time of the Renaissance, the bogeyman had come to be seen as harmless and even admirable, and he had been relocated. In 1501–2, Amerigo Vespucci (1454–1512) explored part of the coast of Brazil and wrote a description of its natives to his prince, Piero Lorenzo de Medici. This was published in Latin, Italian, German and Dutch and became a best seller and very influential.[12] Raphael Hythloday, the fictional storyteller of the original Utopia, sailed with Vespucci and described a people, in fact highly civilized, who disdained possessions and lived in South America.[13]

Vespucci sensationalized the people and customs he found in Brazil by stressing everything that he considered outlandish. The people lived, he said, without money, property or trade, they were completely free, morally and politically, they had no kings or religion and they lived to a great age.

Europeans were enchanted by this life of innocence and, before long, natives were being brought back for display, sometimes in settings meant to recreate their Brazilian surroundings. The Indians were, of course, completely naked. The image made so great an impression that, for a very long time, travellers to South America saw everything in this light. They saw what they wanted to see. Today, our concern for the tribal peoples of South America remains strongly tinged with that image of the Brazilian noble savages, though South America is not the only region in which such paragons have been found. In the 1960s, anthropologists found another group, the !Kung, living in the Kalahari Desert.

Calculations based on a three-week study of the !Kung economy led the scientists to conclude that these people obtained all the food they required by foraging for two or three hours a day and that they enjoyed a varied and nutritious diet. Their society was egalitarian and its members lived in peace. The !Kung were said to be 'among the few remaining representatives of a way of life that was, until 10,000 years ago, a human universal.'[14] The news spread like wildfire among those opposed to industrial society. 'Is it so paradoxical to contend that hunters have affluent economies, their absolute poverty notwithstanding?' asked Marshall Sahlins, the principal proponent of this view. 'Modern capitalist societies, however richly endowed, dedicate themselves to the proposition of scarcity. Inadequacy of economic means is the first principle of the world's wealthiest peoples.' In contrast, he maintained, 'primitive' people live a leisurely, peaceful existence, their wants amply supplied by the plants and animals around them.[15] The hunter-gatherers began to be seen as protectors of a healthy, peaceful, leisured way of life that was better adapted than ours to the natural environment, and that is how many people regard them still.

However, anthropologists now realize that their ideas were mistaken. In the last century, the !Kung were deeply enmeshed in the world economy, first trading cattle and then supplying the market for ivory and ostrich feathers, which they traded for agricultural and manufactured goods. They hunted elephants with guns. A century later, most of the elephants had been killed, the cattle had succumbed to rinderpest, and the fashion for ostrich feathers had died. The !Kung were living in abject poverty, first working for nearby farmers and then being driven deeper into the social underclass and a life based on foraging. By pure coincidence, the three-week study found them enjoying a brief spell of plenty, but usually their life is far from pleasant. Their average life expectancy at birth is 30 years, infant mortality is high and they lose weight alarmingly when food is scarce. What applies to the Kalahari tribes is also true of all the other tribal societies. They are survivors, hanging on in a harsh life to which they have been driven. Ancestors of the forest-dwellers of South America farmed the fertile valleys. Invasions by Europeans and more aggressive tribes drove them into the forest.

Despite having been thoroughly exposed, the myth refuses to die. In 1972, journalists were introduced to the Tasaday, a tribe first encountered not long before in Mindanao, in the Philippines. These were genuine noble savages, or so it was thought. They wore garments made from leaves and lived in caves, and their tools were of stone and bamboo. They made fire with fire drills and ate an assortment of roots, fruits, insects, frogs and crabs. They made no cloth or pottery, had no knowledge of metals, no method for counting the passage of time, and the idea of corporal punishment was quite alien to them. Their language had no word for 'war'. The most primitive people on Earth, the Tasaday lived in idyllic harmony with nature and one another. In 1986, Oswald Iten, a Swiss journalist, visited them again. He found tribesmen who told him they had been paid to remove their ordinary clothes, sleep in caves rather than their thatched huts, eat wild food and swing from trees now and then, and that before visitors were allowed into their area, someone would fly in to make sure everything was ready for the show.

The Tasaday could not have lived isolated in Mindanao since the dawn of time. They were Malay people, like most Filipinos, whose ancestors migrated to the islands about two thousand years ago. However, later studies argued they were not quite so fraudulent as they seemed: linguistic analysis suggested they had been isolated for several centuries.[16] We are left with two alternatives: either they are completely fraudulent, or they are genuine, but have been exploited by being represented as much more 'primitive' than they really are. In either case, the trick worked because of our gullibility. We long to find the noble savages for which we have been searching all these centuries.

Does it matter that we treasure a belief in myths of a Golden Age and noble savages? Yes, it does. In the first place, it is better to face reality than to hide from it behind fantasies. More seriously, if we are sincere in our wish to help others and protect the weak from exploitation, then we must see them as they really are. Otherwise, if we persist in regarding a fictionalized tribal life as preferable to our own, we may well consign real people, living in real poverty with real problems, to a kind of museum where they serve to provide amusement and spiritual uplift to wealthy tourists. Similarly, if we wish to improve the world we live in, we cannot do so by seeking to recreate a Golden Age that never existed. Some myths are helpful and deserve to be retained. Others are harmful and should be abandoned. The Golden Age and the noble savage are of the second type.

10

Mother Earth, Wilderness and El Dorado

Myths are of use only if they relate us to the mysterious. They contain a kind of truth, but to remain a truth the myth must be honest. If the relationship it describes is uncomfortable, dangerous even, the myth must make this clear. If we revise the story to make ourselves feel more secure than is warranted or support an image of the universe that conflicts with evidence obtained from observation, then we distort the myth. The result is a soft, sentimentalized travesty of the original with neither the strength to challenge us nor the wit to inform us. The widespread belief in, and sometimes worship of, 'Mother Earth' is an example. This distorts not only an ancient myth, but also a scientific hypothesis. Indeed, it is a most curious conflation of concocted religion and pseudo-science derived from real science.

Throughout history, people have had a mystical relationship with the Earth, understood as both the soil that feeds us and the planet as a whole. Most of our old stories describing this relationship are Indo-European and they have been told everywhere from Egypt to Scandinavia. It may be an Egyptian version of one of these stories that Orpheus, or some of his followers, the Orphics, encountered in Crete and transmitted to Greece. It tells of a time when Nyx, a great, black-winged bird, was the only being. It laid an egg, which hatched to release Eros, who flew away on golden wings. The two halves of the eggshell then sprang apart. One became Uranus, the sky, the other Gaia, the Earth. Gaia, sometimes spelled as Gaea or Ge, was not a god; the gods came later and this is one of the many creation myths. The name, as Ge, lives on in our language, as a prefix in words relating to the Earth, such as ge-ography, ge-odesy, ge-ometry and ge-ology.

The Gaia hypothesis

In the 1960s, some thousands of years later, the novelist William Golding suggested the name 'Gaia' to his friend James Lovelock as an evocative label to describe Lovelock's new idea about the way our planet functions. Lovelock thought it apt and adopted it. As interest in his hypothesis grew, so did enthusiasm for it among some (but by no means all) environmentalists. From them, ideas about it spread to other groups, inspiring the revival of religious ideas based on an 'Earth Mother'.

The substitution of a female deity for the traditional male god of a male-dominated society has obvious social and political significance and in this context the metaphor is very valuable. But some people have taken it much further. 'Gaia' has become the source of an all-embracing, maternal cosiness, a goddess in whose arms we may happily be nourished and lulled into the deep slumber of complacency, provided that before we doze off we denounce all the works of men as evil. 'The ways of Gaea were forgotten,' Kirkpatrick Sale complains. 'Over the years the Mycenaeans systematically cut down the holly, cypress, olive, pine, and sycamore trees that originally covered the Mediterranean slopes, using the wood for fuel and lumber . . . Herding of goats, cattle and swine as well as sheep became a common practice, quite heedless of the multiple effects on the countryside'.[1] The immediate effects, of course, were to supply food and living space for humans and it is not certain that 'Gaea' ever warned these early farmers of consequences that would appear centuries later. We may wonder whether it is they who forgot 'the ways of Gaea' or Mr Sale who is criticizing them in the light of knowledge they could not possibly have possessed.

In 1985, Joss Pearson conceived and organized the publication of *The Gaia Atlas of Planet Management*, edited by Norman Myers, a well known scientific conservationist.[2] The book was so successful that she was able to set up her own publishing house, Gaia Books, issuing about ten titles a year on ecology, health, natural living and the mind.[3] She sees Gaia as 'a life-style, a marriage of ecologists and conservationists to holistic healers and spiritualists', citing support from readers for whom 'Gaia, the ultimate embraceable whole, had changed their lives'.[4] There is even a hymn, called the Gaia Song, to be sung during a ceremony meant to help Gaia through the dark days of winter.

It is all far removed from the original idea and, while welcoming some of the support it has given him, James Lovelock views the New Age interpretations with irritation. The development of his hypothesis began when he was working for NASA, then preparing for the first unmanned soft landings on

Mars and considering how they might discover whether or not the planet supported living organisms. In discussions with his colleagues, Lovelock came to realize that, regardless of their appearance or way of life, all living beings have one thing in common: they modify their surroundings by taking nutrient substances from them and returning their metabolic by-products to them. It should be possible, therefore, to recognize the presence of life through perturbations to the chemistry of the planet, and in particular of its atmosphere, which is where evidence of perturbation is most likely to be found. These ideas were first published in a few scientific journals and they reached a wider public when Lovelock outlined them in a popular book, published in 1979.[5]

Applied to Earth, the only planet known to support life, the Gaia hypothesis can explain the chemical composition of the atmosphere and oceans, proposing that the biogeochemical cycles, by which elements move between land, sea and air, are driven by biological processes. Thus far the idea is not in the least controversial. It fills in some missing details, though scientists have known for many years that living organisms play a crucial role in these cycles. The difference is mainly one of emphasis. In the Lovelock scenario the living organisms are more than crucial. They are in full control, even to the extent of regulating the salinity of sea water and, possibly, contributing to the tectonic movement of crustal plates. Most important, perhaps, in the light of present concerns, living organisms stabilize global climates by regulating the atmospheric content of carbon dioxide.

As Lovelock suggested, the Gaia hypothesis can exert a powerful influence on the way we think about the natural world and it generates a rich supply of ideas. He and I explored some of these in considering, before it was accepted as widely as it is now, the Gaian implications of the planetismal collision which may have caused the mass extinction that ended the Cretaceous Period some 65 million years ago.[6] We also proposed a Gaian scheme for rendering Mars habitable, which aroused considerable scientific interest.[7] In the end, however, a scientific hypothesis stands or falls not on its entertainment value, but on its explanatory power. Lovelock has used Gaia to explain, for example, the formation of clouds over the open sea by the condensation of water vapour onto sulphate particles derived from the oxidation of sulphur detached from dimethyl sulphide, which is released by several species of phytoplankton.[8] This is one mechanism by which living organisms affect climate.

The interest of scientists should not be misconstrued as their general support. The Gaia hypothesis remains highly controversial. As Peter Westbroek, a geochemist, said, 'life represents a geological force and has deeply influenced the history of this planet . . . But for Lovelock, Gaia is the master of all geological forces, and is in charge of the planet. Which professor of earth sciences would dare to make such a statement!'[9] Gaia is criticized for being circular: the existence of a hospitable environment proves the existence of Gaia and the existence of Gaia explains the existence of a hospitable environment. It is said to explain too much: no matter what the phenomenon, Gaian theory can find an explanation for it. Finally, it is said to be teleological, in supposing some kind of desirable outcome towards which organisms conspire.

The charge of teleology is grave. Humans, and perhaps a few other mammalian species, can make plans for the future and take steps to realize them, but to suggest the collectivity of organisms as a whole can do so is irrational. When you or I make plans and act on them, our idea of the future influences our present behaviour. To suppose that the global climate, say, is maintained in a fairly stable condition by organisms acting in their own interests, however, is to imply either that collectively they have some concept of those interests and means of collaborating, or that the future is influencing the present and, therefore, that effects are controlling their own causes and causality is running backwards. Not surprisingly, this is the charge that Lovelock found most wounding and the one that he devoted considerable effort to refuting. He did so by devising computer models, of what he called 'daisy worlds', to demonstrate that climate regulation can be the inevitable consequence of a purely automatic response to present conditions.[10]

Some scientists find his 'daisy world' demonstration unconvincing. However, the most interesting point about it is that it reveals the nature of the Gaian hypothesis, which Lovelock himself has emphasized repeatedly but which is often overlooked. 'Gaia' is nothing more nor less than a homoeostatic system, a machine. Its self-regulatory responses are purely automatic feedback mechanisms resulting from such ordinary activities as respiration and the acquisition and metabolization of food. It leads to a concept of the Earth as, in some sense, a unified living organism, but it implies no guiding intelligence, no conscious pursuit of goals. 'Gaia' is not an intelligent being and the hypothesis is essentially mechanistic and reductionist. All the more surprising that it should have been seized upon so fervently by those who denounce science precisely because they consider it to be mechanistic and reductionist.

There is a further implication of the Gaia hypothesis: it contains no place for large animals, such as humans. The cycling of elements is mainly the business of bacteria while the regulation of climate is due to single-celled organisms and, to a lesser extent, some aquatic invertebrates, aided by larger plants. Were humans, cows, tigers and whales to vanish, Gaian theory holds that the planet as a whole would be little affected and that small effect would be beneficial. History has shown 'Gaia' to possess remarkable resilience. The system has recovered from extremely violent assaults, such as the late Cretaceous planetismal impact, and grave 'illnesses', such as resulted from its acquisition of an oxidizing atmosphere. 'Gaia' may well shrug off such insults as industrial pollution and even climatic modification, and if those events were to lead to the loss of humans and other large animals, that would be insignificant. The planet would survive.

Much depends on interpretation, of course, and there are 'weak' and 'strong' versions of Gaia. The weak version merely notes that planetary homoeostasis is more directly mediated by living organisms than might have been supposed. The strong version holds that the planet is, literally, a single living organism and it is this interpretation that allows room for further evolution involving humans. Peter Russell, for example, has speculated that human development can be seen as the establishment of networks of increasingly rapid and efficient communications that have brought us close to the point where their complexity becomes comparable to that of a brain. When, in the very near future, that threshold is crossed, the Earth, or Gaia, will attain self-consciousness and our own evolution will enter a qualitatively new phase.[11] It is an interesting metaphysical idea, for which Russell advances a thoughtful argument, but the wider attraction of 'strong Gaia' derives from its aspect as 'Mother Earth', which some people find comforting.

Even mythologically, Gaia can afford us little comfort, for there was a dark side to every tradition describing an Earth Mother. The Earth that sustains us also devours us and when offended the 'Mother' is quick to wreak terrible revenge. The more extreme environmentalists are aware of this, of course, and use it to chastise us for our 'ecological' sins. Nevertheless, the tautology that terrestrial species can survive only on Earth is popular and those who think it profound find support for it in Gaian theory. It makes the Earth sound cosy, our home, 'spaceship Earth' that shelters and protects us from the harshness of the extraterrestrial environment and which, in our own interests, we must safeguard. The cosiness is closely linked with a popular and highly selective view of non-human animals. These, too, must be protected, most of all against the risk of extinction.

Conservation and sentimentality

Most of us would agree, I am sure, that the conservation of species is desirable, but not perhaps at any price and, for some people, not of all species. I remember asking, when I was a small boy, what possible use God could have had for wasps (which frightened me because they sting). Would those of us who favour conservation extend our protection to all species, including, say, *Mycobacterium tuberculosis*, which causes tuberculosis, or the mosquitoes which transmit malaria and other diseases? Would we preserve the human louse? Were scientists to announce a means for bringing to extinction the human immunodeficiency virus (HIV) would we conservationists petition to restrain them? We are selective, even in our attitude to larger animals. How fond are you of slugs? Do you really love big spiders?

Our efforts are concentrated on those species most like ourselves or those we can think of as being like ourselves, even if this requires us virtually to reinvent the animal. Whales, for example, are so popular as to have acquired a quite mythological status. Some species, most notably the humpback (*Megaptera novaengliae*), communicate by means of an elaborate 'song'. This is taken as evidence of intelligence, though the communication seems to amount to no more than a mating call. The large baleen whales are about as intelligent as cattle. The smaller, toothed whales and porpoises are probably much more intelligent and comparable to dogs, but not to humans. Elephants are also popular, especially among people who have not suffered the destruction of their crops by trampling herds, and who have difficulty accepting the message from professional conservationists, that the total prohibition of hunting them for ivory has made control much more difficult and endangered them by increasing the demand on their limited food supply as well as increasing the risk of disease among densely crowded populations.

Sentimentalizing animals does no favours even to those we wish to protect and it ignores the needs of those it is difficult to think of as 'cuddly'. Further, should it turn out that the planet would benefit from our protection, we will not protect it by regarding it as an intelligent, benign 'mother'. Conservation and environmental protection are ill-served by such anti-science and pseudo-science. These desirable objectives can be attained only through rigorous scientific observation, experimentation and debate and it is reasonable to ask whether those who reject the means genuinely will the end. The question is a serious one, because both means and ends have implications. Unless we believe them to be attainable, it is pointless to attempt them. If we believe them attainable and attempt them, we do so in the wider belief that it is possible to progress towards a future that is better than the present. We assume that past errors can be remedied, that wrongs can be redressed, and that we are capable of learning from our mistakes so as to avoid repeating them.

In other words, acceptance of the aims of conservation and environmental protection, and of the scientific means of achieving this, perhaps informed by Gaian theory, supposes a belief in the future. The opposite view, by rejecting the means, makes the end unattainable but leaves unaltered, except by exaggeration, the scientific diagnosis of what needs to be done. Since doing what needs to be done is now impossible, the future will necessarily be worse than the past or the present, so confirming our decline. The many who hold this view are unable to accept the feasibility of a benign kind of progress and are reduced to bewailing the loss of the cosy world and cuddly animals of their imagination. The kindest description of their attitude is 'childish'. A more honest one is 'decadent'.

Animals and tribal peoples, being nice and hence deserving of our protection, must be provided with suitable land where they can live at peace and in harmony with their natural environment. Free from human management or, ideally, intervention of any kind, such lands are classed as wilderness and we prize them highly. Wilderness is usually defined as an extensive area that has never been permanently occupied or used intensively by humans and is in a natural, or almost natural condition. The definition centres on the ecological integrity of the area rather than any particular feature it may possess. In the United States, wilderness areas are designated formally and they can be put to economic use only by presidential decree and then only in an emergency. Traffic is not allowed and the activities of visitors are controlled.

In Europe, where there are few tracts of virgin land, of necessity the definition is wider. In Britain, for example, the concept of wilderness merges imperceptibly with that of the national park. British national parks are occupied and most of their area is farmed. The 'national park' designation is intended to protect land from changes in use that would significantly alter its character. Areas of valuable natural habitat more closely conforming to the American idea of wilderness are mostly small, and more often protected within nature reserves, where public access is controlled and in some cases forbidden.

Wilderness areas have not always been regarded as an asset. They used to be called 'wastelands' or simply 'waste', the uncultivated area that lay beyond the boundaries of the farms. Often they were forested and used mainly for hunting game. Elsewhere they were barren mountains or deserts. Those who ventured into the wilderness risked death from hunger, thirst, cold, heat or encounters with wild beasts. It was a place to be feared. Early European explorers in North America paid close attention to the cultivability of the lands they saw, pronouncing most of them satisfactory. But when the first Puritan settlers arrived and saw the vast forest, they described it as 'heidious and desolate wilderness full of wilde beasts and wilde men.'[12]

Their attitude is understandable. When famine was a familiar hazard, the best guarantee of security lay in the cultivation of the largest possible area. Hampstead Heath was described by John Houghton in 1681 as 'a barren wilderness',[13] reflecting the general opinion that well farmed land was the hallmark of civilization. In the early nineteenth century, Arthur Young, commissioned by the English Board of Agriculture to report on progress made in enclosing open lands to make them agriculturally more productive, complained of the slow rate at which 'wastes' were being brought into cultivation. In Needwood Forest, he wrote, 'near 10,000 acres of one of the finest soils of the kingdom lie in a state of nature'. In Lancashire he found moors, marshes, commons and salt marshes capable of improvement. 'Why seek out distant countries to cultivate, while so much remains to be done at home?' he asked.[14] A few years later, William Cobbett described the countryside that he deemed 'the brightest and most beautiful and, of its extent, best of all' he had seen on his travels:

> Smooth and verdant down in hills and valleys of endless variety as to height and depth and shape; rich cornland, unencumbered by fences; meadows in due proportion, and those watered at pleasure; and, lastly, the homesteads, and villages, sheltered in winter and shaded in summer by lofty and beautiful trees; to which may be added, roads never dirty and a stream never dry.[15]

A change of attitude had been in the air, however, as early as the middle of the eighteenth century. Without mountains there would be no rivers to water the lowland fields and ranges of hills formed convenient boundaries. Quite apart from this utility, some people thought that denial of the beauty of the more rugged landscapes implied distrust in the perfection of creation. Wastelands began to acquire an aesthetic appeal that they have retained ever since. As educated people took up botany, hillwalking, mountaineering, or simply appreciating the scenic beauty, visits to the awesomely wild places increased, roads and transport improved to meet the demand and tourism began.

While Europeans were adopting a religious view of unspoiled landscapes, basing their aesthetic appreciation on the almost mystical qualities they attributed to rugged nature, another, more scientific approach to conservation began to exert an influence. Its instigators were also European, but their concern centred not on Europe or North America, but on tropical landscapes, and their concerns have a distinctly modern resonance. One book, above all, captured the imaginations of its readers and launched the scientific discipline of biogeography. *Ideen zu einer Physiognomik der Gewächse* ('Ideas towards a physiognomy of plants') by Alexander von Humboldt (1769–1859) was published in 1806. (Physiognomy is the structure and form of plant communities.) In 1800 von Humboldt, a mining engineer, geographer and geologist, had spent four months in the company of the French botanist Aimé Bonpland exploring the Orinoco and Amazon basins. Later, after a short break in Cuba, they visited the sources of the Amazon and explored the Pacific regions of South America. From his studies of the plant and animal life of the regions he visited, von Humboldt devised a concept of the interrelatedness of people and the natural world.

The relevance of this concept was recognized very quickly by scientists attached to companies in Dutch, French and British colonies, many of whom were physicians doubling as curators of botanical gardens. As the companies expanded into new areas, so the number of supporting doctor–scientists increased. In 1838 the British East India Company was employing more than 800 surgeons in India and the East Indies.[16] Eventually, scientific societies were formed, whose members expressed a growing concern about the extent to which natural environments were being altered through exploitation. Word of these concerns soon reached Europe and, in 1850, the British Association for the Advancement of Science commissioned a study to consider 'the probable Effects in an Economical and Physical Point of View of the Destruction of Tropical Forests'. The resulting report, by Edward Balfour, Alexander Gibson and Hugh F.C. Cleghorn, of the Madras Literary and Scientific Society, was published in 1851. It strongly recommended conservation and led directly to the establishment, in 1855, of the Indian Forest Service, employing officials called 'conservators' whose task it was to achieve sustainable forestry yields.[17] The Indian service provided a model for other colonial forestry services to follow, but some steps had been taken even earlier. French forestry conservation measures, enacted for Mauritius in 1769 and 1803, led the British to create forest reserves in Tobago in 1764 and St Vincent in 1791. Reflecting yet another supposedly modern concern, the French conservation measures in Mauritius and British ones in South Africa were motivated partly by fears that deforestation would produce adverse climatic effects.

Employees of the colonial services were practical people in daily contact with the lands they administered, who acquired the scientific understanding that allowed them to diagnose environmental ailments and prescribe conservation remedies. In Europe, changing environmental attitudes were driven largely by religious and aesthetic considerations. There was a third, more complex strand to this change, however, intricately interwoven with the others and as influential today as ever it was.

The myth of El Dorado

Few of us are immune from nostalgia, though the emotion is stronger in some parts of the world than in others. The Germans express it in their word *Heimat*. It means 'home' but, much more than that, a sense of the place from which you come and the thousands of years over which the ancestors of your family and community have toiled to shape it. It is the place for which we pine when we are away, to which we are always drawn by ties rendered no weaker by the difficulty we may experience in trying to describe or explain them. Mix with this feeling a mystical longing to see the world as it was when first created, when it was cleaner, more wholesome, than it has been left by the corruption with which we have infected it, and the idea of the pristine wilderness seems eminently worth protecting. This feeling is powerful and few of us can escape it. But along with it should go a 'health warning' to the effect that this wilderness lives almost exclusively in our minds. In the real world, plant and animal communities are a deal more dynamic than the romantic image suggests. They change over time, and are subject to natural disasters such as floods, gales, landslides and fires, while the 'balance of nature' some of us treasure so dearly does not exist. It, too, is a myth.

We can and do try to create our own versions of paradise. Gardens are no more 'natural' than city blocks, but we make of them small oases of beauty and tranquillity, our private paradises. Some Europeans prefer more formal-looking gardens whereas the English have a fondness for 'cottage' gardens, made in the image of a fancied rural idyll. Real country cottages, in which farming families live, usually have gardens filled with vegetables, chickens and washing, but the riotously coloured flower-beds and climbing roses around the door of the invented cottage garden are more pleasing reflections of the beneficence of nature. Americans, no less susceptible to nostalgia, often

107

reflect 'a romantic longing for the pastoral life-style of the pre-Civil War period, even though this style did not last long and was always limited to a few regions and people'.[18]

Reports from the explorers of exotic lands often described idyllic conditions. 'The soil there is so fertile,' wrote Father Jérôme Lalemant, in 1661–2, of land probably in the Ohio valley, 'that one could almost say of it, within bounds, what the Israelite discoverers said of the Promised land; for, to mention the Indian corn only, it puts forth a stalk of such extraordinary thickness and height that one would take it for a tree, while it bears ears two feet long with grains that resemble in size our large Muscatel grapes'.[19] Henri Joutel visited the bay of St Louis in 1685 and found it a veritable Eden. After listing the abundant mammal, bird and fish life, almost all of it deliciously edible, and the vines and mulberry trees, he concluded: 'Nothing is more beautiful than to behold those vast plains when the blossoms appear. I have observed some that smelt like a tuberose, but the leaf resembles our borage. I have seen primroses having a scent like ours, African gilliflowers, and a sort of purple wind flowers'.[20] Their imaginations fed by hundreds, perhaps thousands, of reports from different authors but all couched in such glowing terms, who can blame Europeans for picturing a New World flowing with milk and honey, simply waiting to feed them?

That world might also contain immense riches. Somewhere east of Quito, stories told of a king so wealthy that once a year he took part in a ceremony, held in the middle of a lake, in which he was covered in gold dust. He was the gilded man, El Dorado. There was another land of fabulous wealth, in this case rich in cinnamon, La Canela, and in 1541 Gonzalo Pizarro led an expedition to discover both.[21] He failed, but other expeditions searched in other places. Walter Ralegh led one of them, in 1595. He did not find the fabled lands, of course, for they did not exist, but of the region he explored he wrote that

the common soldier shall fight here for gold, and pay himself, instead of pence, with plates of half a foot broad, whereas he breaketh his bones in other wars for provender and penury . . . Guiana is a country that hath yet her maidenhead, never sacked, turned, nor wrought, the face of the earth hath not been torn, nor the virtue and salt of the soil spent by manurance, the graves not been opened for gold, the mines not broken with sledges, nor their images pulled down out of their temples.

As John Hemming wryly comments, Ralegh omitted just one point: the place he described had not yet been discovered.[22] His story, like the others, was a fiction born of wishful thinking.

Today, of course, we would be appalled at the very idea of such plunder. Ravening greed may characterize multinational corporations but it has no place in the value systems of those of us sensitive to the needs of indigenous peoples and their natural environment. We might accept that, centuries ago, Europeans justified their behaviour by their sincere belief that they were conveying the true faith to savages who would otherwise be denied even the possibility of paradise, and they had to be paid for their work, but even this would be regarded today as a form of imperialism. We are no more entitled to impose our beliefs on others than we are to rob them of their resources. At least, that is what we choose to think. But consider the reasons we are given for preserving tropical forests.

A question of motive

We are told they contain half or more than half of all the species on Earth and that these must not be permitted to become extinct. However, their preservation is urged only partly on moral or aesthetic grounds. We are also told that, among the plants especially, there may well be those from which we could obtain drugs to cure many presently intractable ailments. Others might be exploited as raw materials for new manufacturing processes. Economically, the primary beneficiaries of such exploitation would be the inhabitants of the countries in which the forests occur, but it is our illnesses that may be cured and it sounds like economic exploitation to me. Pharmaceutical products are now worth more than their weight in gold, which makes the tropical forests seem curiously like the kingdom of a modern El Dorado. In pursuit of this self-evidently laudable objective we are perfectly happy to instruct those inhabitants and their governments in the correct management of their forests.

Given that, among us, conservation and environmental protection have assumed something of the status of a religion, our efforts do not seem too dissimilar from those of our ancestors. Like them, we are seeking to impose on others our own beliefs and, also like them, our aim in doing so is to bring to the unenlightened the glad news of the kingdom of ecological heaven, where a place is reserved for those who open their hearts and minds to the creed so generously presented to them.

Not that environmental protection and the conservation of species and habitats are unworthy objectives. It is the attitudes that motivate some of us that are questionable. If these are substantially based on quite venerable myths, the desired protection and conservation may be more difficult to achieve than is supposed. We cannot justify converting vast regions of the world into museums to satisfy our nostalgia. Even less can we hope to recreate that which never existed. If we imagine that wilderness areas contain wealth, in whatever form, we may be deluding ourselves just as earlier explorers were deluded. Our disappointment and frustration at the resulting failure might then lead us to an older image of wilderness. Like our ancestors, we might come to see it as essentially hostile, uninhabitable and fit only to be converted into a more economically productive resource.

Conservation can succeed only if based on sound scientific procedures. If that is what we seek, then we first have to understand those procedures and then adopt them or, if that proves difficult, support those who can do so. The first step in the appraisal of the task before us is to see the world as it really is, and not as we would like it to be.

IV

FAUST AND HIS DEBT

11

Science, Technology and Wealth Creation

What is the most potent weapon ever to have been devised in the whole of human history? The atomic bomb, hydrogen bomb, doomsday bomb or, perhaps, nerve gas or biological weapons? Faced with this question, many people would have chosen one of these, but none of them is correct. The most potent weapon by far is the bow and arrow. It has been used more often, to deadly effect, than all other weapons combined. In medieval England adult males were required to train in its use so that in time of war the king could call on a vast army of skilled archers conscripted by their feudal lords. They used the longbow, with which they could achieve a rapid rate of fire, to devastating effect. The Mongolian armies that swept across Europe also relied on skilled archers, often firing from horseback and using a bow of their own invention with arrows specialized for particular purposes. Both English and Mongolian arrows could pierce armour. In South America, bows were invented that were so large and required so strong a pull the archer had to lie on his back, holding the bow with his feet and drawing the string with both hands. Among some of those tribes it was a mark of shame to fire an arrow that missed its target, as many Spanish and Portuguese soldiers discovered to their cost.

To this day, children play with bows and arrows and archery is a popular sport. Country shows and fairs often have butts as a side show, and they attract queues of adults as well as children. The bow and arrow was invented by different peoples, independently of each other, in many parts of the world. This is not surprising, for the familiarity of the weapon makes it seem simple and obvious. In fact, the simplicity is misleading. It is a subtle device, capable of much modification and sophistication. Apart from its use as a weapon, it has also been used as a drill, to bore holes, and to light fires. Further refinement turned it into the crossbow, capable of imparting a much greater velocity to its projectile but only by sacrificing the high rate of fire of the longbow archer. Crossbows are still used for sport, and also by criminals. They can kill as effectively as many firearms and they are silent. The firing mechanism of the crossbow led eventually to one that could ignite an explosive charge and thus firearms were invented. Even then, the old-fashioned bow retained its supremacy for a very long time.

Science or technology?

The point about the inventors of the bow and arrow, whoever they may have been and whenever they may have lived, is that they did not approach their task scientifically. Neither did their successors who invented the descendants of the bow. Indeed, it is only in the last year or two that scientists have begun to explain just how the weapon works.[1] What is true of the bow and arrow is no less true of countless other devices that have been invented over the centuries. Machines have been invented by engineers, cathedrals and palaces designed by architects and medicines devised by apothecaries, but none of them sought to reveal the scientific principles underlying their products. Sailing ships plied the oceans long before anyone understood the physical laws governing their motion, and they navigated with the help of magnetic compasses centuries before scientists began investigating the geomagnetic field and ferromagnetism. Inventors were simply imaginative people who could visualize what would work and what would not, and were skilled at identifying practical problems for which solutions might be available.

Technology is often described as the application of scientific principles. In a sense this is true, since by definition scientific principles underlie all phenomena but, stated this way, clearly it is a truism. When the word entered our language, around 1615, it meant a discourse (from the Greek *logos*) on an art (Greek *tekhni*) or, more precisely, the (scientific) study of the industrial or practical arts. In other words, inventors studied ways to make things work, and some time later scientists studied why they worked. The technology came first. Indeed, it came first by many thousands of years, for even our probable ancestor *Homo erectus* used tools and weapons, which are technological devices exploiting scientific principles.

In recent years the distinction between science and technology has become blurred, but it is important to bear in mind that there is a distinction, even now. Scientists engaged in 'pure' or 'basic' research are concerned to expand our understanding of the universe and all it contains. It may be that this understanding can be exploited technologically, and if this seems likely then scientists will be recruited to develop applications that could lead to manufacturing processes or products. The blurring of boundaries sometimes creates ethical dilemmas, which must be addressed, but it is still legitimate to distinguish between explanations (science) and devices (technology).

Not all applications of scientific principles are technological, of course. Ecologists study relationships within communities of living organisms. Their work contributes to our general understanding of the world we inhabit, but it can also inform us of ways to avoid harming non-human species, and to provide and manage areas that will support species or communities that otherwise would be at risk. Palaeoclimatologists study the climates of the distant past. This contributes to our knowledge of the history of our planet, but palaeoclimatological data are also used in calculations of the possible future climatic implications of present activities, such as the release into the atmosphere of so-called 'greenhouse' gases. Areologists study the physical structure of the planet Mars, providing information that can also tell us much about our own planet, perhaps helping to guide us in our stewardship of it.

Very often, what purport to be criticisms of the work of scientists concern not the acquisition of understanding, but products or devices and their uses. In other words, they are criticisms of technologies and our sometimes incautious use of them. Such criticisms may be justified, though some fall into the trap of oversimplifying what, in fact, derives from a very complex situation or decision.

Unheeded warnings

I am of the generation of people who reached adulthood under what seemed to be the constant threat of thermonuclear annihilation. Like many others, I marched against 'the bomb' and opposed the atmospheric testing of nuclear weapons. Since then, the enmity between two awesomely powerful political blocs has dissipated, but the world is, if anything, more unstable today than it was before and the threat of nuclear warfare has not disappeared. War could still be unleashed by accident, for example.[2] Should 'the bomb' have been made at all? The question is often asked, despite its irrelevance: it was made and we cannot 'unmake' it. Nor, in any meaningful sense, can we 'ban' it, for knowledge of it cannot be eradicated. In a world in which all such weapons had been verifiably destroyed, in the event of serious conflict the advantage would lie with the combatant who could reconstruct and use it first. A total ban, were it feasible, might well decrease security rather than enhance it.

Physicists recruited during the Second World War to make the first atomic bombs were prominent among those who warned of the dangers that would arise from their proliferation. Issues of whether such weapons should be made and used must be decided by society at large, through their politicians. But, as in all such ethical controversies, the scientists and technologists have much to contribute from their specialist knowledge. In the case of thermo-nuclear weapons, scientists have never ceased campaigning. In December 1983, for example, a group of them published in *Science* the conclusions they had reached about the possible climatic consequences of all-out war in the northern hemisphere, later summarizing these in more popular form.[3] Their scenario became known as 'nuclear winter' and stimulated much discussion and further scientific investigation. Its reception might serve as a warning, however. The authors of the study sought to persuade politicians that the effects of full-scale war would be so disastrous for all concerned as to make it unthinkable that any country could contemplate the use of nuclear weapons. Therefore, they argued, the politicians should publicly eschew such use and dismantle their national armouries. Politicians saw things differently. They agreed that nuclear war was unthinkable and maintained the best means of prevention lay in holding, and if necessary increasing, their arsenals for purposes of deterrence.

There are occasions, however, when politicians and industrialists do respond to scientific advice. As I mentioned in chapter 5, the adverse ecological effects of uncontrolled pesticide use were recognized by British scientists in the 1950s. Their warnings were heeded and the worst dangers were avoided. Far from being the authors of our follies, scientists are more often the guardians of our well-being. It was they who warned us that fallout from atmospheric testing of nuclear weapons might pose a biological hazard, they who identified the mechanisms by which organochlorine insecticides reduced the breeding efficiency of birds of prey, and they whose concern over a very wide range of issues leads them to rigorous investigation to identify possible risks. Heeded or ignored, they cannot be justly accused of not trying or of irresponsible behaviour.

It is not always politicians or manufacturers who ignore scientific advice. Indeed, to do so could sometimes lay them open to charges of recklessness. We, the public, also tend to ignore or dismiss information that we find, or are encouraged to find, inconvenient. The use of food preservatives is a case in point. There are those who urge us to avoid 'additives' of any kind, blaming them for a vast array of alarming, but often rather vague, ailments. True, from time to time researchers identify a possible minor risk and a particular substance is withdrawn, though there is no reason to suppose anyone was actually harmed by it. But, in general preservatives greatly increase the safety of the food that city-dwellers consume. Those who loathe what they call 'chemicals' overlook the fact that without them we could be at risk from a great variety of types of food poisoning.

Such anti-scientific views can exert great influence and they can be potentially harmful. In 1986, the European Commission issued a directive banning the use of a number of growth-promoting hormones in cattle husbandry. These were hormones produced naturally by cattle themselves, but increasing the dose made the animals produce leaner meat, which consumers demanded. The hormones were administered by means of an implant in the ear, a part of the carcase that cannot come into direct contact with meat intended for consumption. Concerned that this practice might be unsafe, in 1983 the Commission had requested a thorough investigation by a committee of 22 senior endocrinologists and toxicologists. They completed their study, but just four days before they were due to report back, the Agriculture Commissioner, Frans Andriesson, suspended the committee, effectively gagging them, and use of the hormones was banned. In fact, their findings showed that the procedure presented no danger whatever to human health, not even to very young children, considered to be the group most at risk.

You might think this a matter of small import, since only the profits of farmers and the meat industries were likely to be affected. Unfortunately, it is not so simple. Consumers prefer lean meat, farmers can produce it more easily by using hormones, and an ear implant is easily detected. It would be perfectly easy to implant the hormones elsewhere in the body, say beneath the skin, where the implant would be almost impossible to find. Should farmers do so, hormone residues might well remain in meat sold at retail outlets, but only in some meat, unpredictably, and, for practical purposes, it would be undetectable. The hormones themselves have other uses and so are not difficult to obtain, albeit unlawfully. Exaggerated fears and the rejection of scientific findings may well result in the very harm campaigners sought to avoid.[4]

Vested interests

Public distrust of scientists stems in part from the blurring of boundaries between science and technology, between discovery and manufacture. Most governments, perhaps all governments, justify public expenditure on scientific research in terms of the economic benefits the scientific enterprise has brought in the past and will bring in the future. Politicians remind their electorates of the splendid machines 'our scientists' have invented, the new drugs to relieve old ailments, and the new surgical equipment and techniques by which previously intractable conditions may now be treated and lives saved. At the same time, the politicians demand of scientists that they tailor their research to 'economic needs', that they award a higher priority to research proposals that are 'near the market' and can be translated into the greatest return on investment in the shortest time. Dependent, as they are, on politicians for much of their funding, scientists have little choice but to comply. Like the rest of us, they are members of a society that rates the creation of wealth as the greatest possible good. Many have reservations, but keep them to themselves in what they perceive as a climate hostile to the pursuit of understanding for its own sake and the idea of an inquiring, creative spirit.

In such circumstances no one should be too hard on people who are suspicious of conflicts of interest. When we learn that the distinguished professor assuring us of the safety of a particular product holds a consultancy with the company making it, we cannot be blamed for wondering whether his fee might conceivably cloud his professional judgement. Even if the professor holds no consultancy with any firm, some people may still distrust him because of his association with those who do, or at least wonder about the source of some of his research funding.

This attitude can have damaging effects. It impugns the integrity of individuals working in a profession that prizes intellectual honesty as the supreme virtue, and plays into the hands of those who would like to discredit scientists by representing them as venal. This makes it easier to dismiss all scientific pronouncements, but especially those made by the scientists who present themselves as 'experts'. The scientist most likely to understand the safety of a nuclear reactor, for example, is a nuclear engineer, and a nuclear engineer is most likely to be employed by the nuclear industry. If a nuclear engineer declares that a reactor is unsafe, we believe him, because clearly it is not to his advantage to lie about it. If he tells us it is safe, on the other hand, we distrust him, because he may well be protecting the employer who pays his salary. Taking this argument to its extreme, we should believe only those who are disinterested, which means that sometimes we will find ourselves trusting people who are unqualified to advise us. Moreover, an apparently 'disinterested' scientist who may have no commercial connection with the activity being debated, might have other relevant connections. An independent scientist who, for example, informs us of a pollution hazard might be speaking on behalf of a political pressure group dedicated to the closure of a certain or to discrediting the entire industry. This situation is not merely hypothetical, of course. Supposedly disinterested experts do appear on our television screens whenever an even remotely environmental controversy arises. On the face of it, they issue honest warnings and advice, but might all this not be political propaganda that we fail to recognize for what it is?

Confusion of science with technology and the strong emphasis on the commercial and industrial benefits to be derived from research lead into an ethical morass from which there is no obvious escape route. Probably, the best course is to try to remember at all times that, despite their conflation, scientific and technological enterprises are distinct. We can trust knowledge and understanding, so long as we also remember its provisional nature. Scientists must, if necessary, be permitted, indeed encouraged, to change their minds. In matters relating to technology, we must use judgement. This means asking questions. Does the person offering advice have special knowledge of the topic that would make the advice worth accepting, or is he or she a deal less 'expert' than we are led to suppose? If we accept that the person is qualified to advise, then is the advice biased, for example by support for commercial enterprise or a consumer, environmentalist or other pressure group? In a word, do we believe our informant to be honest?

Each of us must find our own answers and in the end there is only one way of doing that with confidence. If we are concerned about an issue, we owe it to ourselves and to society in general to make the effort to understand the scientific arguments that underlie it. Only then will we be able to weigh one view against another and reach an informed opinion of our own. You might call it 'democracy', a system that can work as it should only if each individual is adequately informed of the issues that are to be decided, and fully understands their implications.

12

Reductionism, Holism and the Curious Nature of Truth

'I use the word "reductionism" here broadly to designate that peculiar sensibility which degrades what it studies by depriving its subject of charm autonomy, dignity, mystery.' Thus wrote Theodore Roszak some twenty years ago.[1] He completed his definition by quoting Kathleen Raine's. It is, he said, 'the style of mind which would have us "see in the pearl nothing but disease of the oyster"'. (With hindsight, we may think the image unfortunate, since many people who today attack the alleged reductionism of the scientific method are also sympathetic to the sufferings of non-humans and might indeed link the pearl decorating the throat of a socialite with the disease of the animal from which it was taken.)

Essentially, though, the accusation stands and is never questioned by anti-scientists, or even by scientists when they stray outside their own disciplines Physicist Danah Zohar and Ian Marshall refer to 'Darwin's mechanistic and reductionist theory of evolution', for example.[2] Scientists stand charged with studying phenomena by means of dissection, of 'taking things apart' examining each component in great detail, and then assuming that the phenomenon as a whole can be understand by extrapolation from what is known of its components. In other words, scientists allegedly assume that phenomena, such as complex ecosystems or human beings, are nothing more than the sum of their constituent parts. It is this 'nothing more than' that gave it the name 'reductionism'. Biological phenomena, for example, are supposedly reduced to nothing more than physical and chemical interaction among atoms and molecules.

Holism vs. reductionism

That reductionism has led to increasing specialization among scientists is a further complaint. This is perfectly true so far as it goes, and inevitable because of the sheer volume of detailed information it has yielded. 'In a world of experts,' asks Roszak, 'what becomes of the imaginative energies of ordinary people? . . . what do they know about anything that some expert does not know better?'[3] This is a reasonable question, but one that might be directed to many fields of learning besides the sciences. How well do 'ordinary people' understand modern theories about art, literature or music? It is curiously selective to castigate the 'philistine' who stares uncomprehendingly at an abstract painting or prefers a 'good tune you can whistle' to an atonal composition, while praising the 'common sense' of those who complain of difficulty in understanding scientific concepts. It is hardly the attitude that might be expected to emerge from a scientifically dominated culture.

The remedy for the disease of reductionism is, of course, holism, a word believed to have been coined by the South African statesman, soldier and philosopher Jan Christiaan Smuts (1870–1950). Ecology is perceived by some as a holistic discipline, which explains its popularity among those who apply the name to what, in fact, is partly a religion and partly a political philosophy. Setting this trend, Roszak wrote of ecology as the 'subversive science' whose 'sensibility – wholistic, receptive, trustful, largely non-tampering, deeply grounded in aesthetic intuition – is a radical deviation from traditional science'.[4]

In the twenty years since those words were written the popular image of ecology has continued to diverge from the scientific practice. As a scientific discipline, ecology remains, as in fact it was then, securely located within the framework of 'traditional science' and as reductionist as any other. Bryan Appleyard maintains that ecology grew from 'holistic biology', a systematic, anti-mechanistic view of the world developed in the last century in response to a perceived despoliation of the environment, combined with calculations of the stocks of non-renewable resources.[5] This may sound plausible to environmentalists, but it is quite untrue. In fact, the word 'ecology' was coined in 1866 by the pro-Darwinian zoologist Ernst Haeckel (1834–1919) in his two-volume *Generelle Morphologie der Organismen* and he meant the word to be used scientifically, not philosophically. Sir Arthur Tansley (1871–1955), one of the most influential founders of the discipline, used the word in the sense it has always been used scientifically: 'In its widest meaning ecology is the study of plants and animals *as they exist in their natural homes*; or better, perhaps, the study of their *household affairs*, which is actually a secondary meaning of the Greek word' (his italics).[6] 'Holistic biology' it may be, but it has never had anything whatever to do with environmental despoliation or the depletion of resources by humans.

Appleyard traces the development of reductionism to the emphasis placed by Christians on details of the life of Christ. He points out that: 'Chinese religion was holistic. True knowledge was knowledge of the whole, not the parts. Experiment in the Galilean or Newtonian sense would thus be meaningless.'[7] There may be some truth in this; the Chinese were splendid inventors and engineers who produced a long list of technological innovations, and for many centuries Chinese astronomers were far in advance of their European counterparts. Nevertheless, the Chinese developed nothing we would recognize as 'science'; their vision of the universe was essentially magical, as was that of all the early civilizations with the single exception of Greece.[8]

Once it is accepted that reductionism is wrong-headed, not to say spiritually demeaning, any statement made by someone labelled a 'reductionist' can be dismissed as meaningless. The word has become a term of abuse, and whenever this happens there is a good chance that the definition has lost most of its original meaning. The 'reductionism' of anti-scientists is a caricature and about as helpful as 'fascist scum' or 'commie bastard' applied to people whom we do not like or disagree with politically. By the same token, the 'holism' we equate with epistemological virtue is no more than a vague, usually unhelpful, and quite unattainable ideal, except in the field of quantum theory. There, it has a very special and restricted meaning related to the coordinated behaviour of pairs of particles that have been separated. It is not at all what anti-scientists think they mean when they use the word.

In defence of reductionism

As often happens, the confusion arises from our tendency to muddle hierarchical categories of explanation. The muddle derives from the logically transitive character of explanations. This holds that, if A can be explained in terms of B and B can be explained in terms of C, then A can be explained in terms of C. While true for simple propositions, it requires modification where A, B and C form a hierarchy, with B and A each introducing new information that was not present at the lower level: in this case A may be explicable in terms of C, but only partially so.

A physicist, for example, understands much about the structure of atoms and the behaviour of their component particles. A chemist uses some of the understanding acquired by the physicist to develop a further, or higher-level, understanding of the formation of molecules and reactions between atoms and molecules of different species. A biologist finds all of this useful, but is more interested in the way atoms and molecules are organized to make up a living organism, while the concern of an ecologist is the relationships among organisms. 'Each science contains not only the informational content of the sciences below it in the hierarchy but contains specialized notions of its own which do not appear at all at lower levels.'[9]

This increasing richness of explanation that develops up the hierarchy is a phenomenon that arises from the new ideas added at each level to escape what would otherwise be unacceptable restrictions on the interactions being examined. It does not imply that scientists seriously believe that, for example, a living human being, or bacterium for that matter, can be comprehended solely on the basis of the physical character of its constituent particles, or its bodily chemistry.

If all this sounds too abstract, think of the example of the person who calls to repair your television set. The television engineer is a little like a doctor. The consultancy begins with your description of the symptoms. The engineer listens to what you have to say and, then, after a moment for reflection, removes the back of the set, inserts measuring devices and checks components. Eventually, all being well, a faulty component is identified and replaced. Without a thorough understanding of each component and its function in the whole system that is the television set, the engineer would be as helpless as you or me. At the same time, the engineer needs only an approximate idea of the theories underlying electronics, of the quantum events by which semiconductors work, and of the flow of electrons that comprises an electric current.

A television set is a physical device. The physical principles behind its operation are very detailed and precise, but at the same time capable of a very broad application, because they permit the construction not only of television sets, but of the full range of electronic equipment. By itself, however, knowledge of the physical principles is not enough to build a television set. To do that, it is necessary to understand how principles lead to components and how components should be assembled. There is a hierarchy of knowledge, with entirely new items being added at the higher levels, where understanding of the lower levels can be more general. The repair engineer operates at one of those levels. The makers of television programmes

123

operate at a still higher hierarchical level. They need very little knowledge of electronics or of the detailed operation of the equipment they use, but bring a whole new range of understanding and skills to the tasks they perform. The repair engineer, at a lower level, requires no knowledge of how programmes are made.

It may seem that here I have offered a hostage to fortune, having used the metaphor of what undoubtedly is a machine. Critics of reductionism have used the same metaphor, but in a different way. Instead of our repair engineer, they introduce a scientist of the reductionist tendency from some extraterrestrial civilization where television sets are unknown (would-be migrants should be warned that this civilization is as mythical as Ralegh's El Dorado!). Confronted with the television set, which is switched off, our puzzled alien proceeds to dismantle it with true reductionist zeal. By the end of the examination much has been revealed. The alien knows of all the components, the way they are connected, how current flows from here to there, and how electrons are fired at the back of a phosphor-coated screen. Only one thing remains mysterious: what is all this for? What does it actually do? The reductionist, you see, is capable of comprehending the parts, but quite unable to perceive how they fit together as a coherent, functioning whole. That ability is possessed exclusively by holists.

More likely, the reductive alien would discover among the components the on-off switch and try it, whereupon the final secret would be disclosed. The point is not trivial. If this alien were of the holistic persuasion, and equally innocent of the joys of the game shows, would holism reveal the presence and manner of operation of the switch? Would the holist notice that the aerial, disconnected from the back of the set, performs a vital function and should be reconnected – and, if so, to what?

I now reclaim my hostage. The machine provides a useful metaphor for many aspects of the real universe, but at no time does its use imply that anyone believes the universe or any part of it actually is 'nothing but' a machine. If you wish to study human physiology, it makes no sense to pretend that our bodies do not contain components. If you want to help the sick, you might find it very useful to consider the location, structure and function of the heart, lungs, liver, kidneys and all the rest. Were you the victim of, say, a renal disorder, you might prefer to visit a physician who, you understood, knew where your kidneys were located and what function they performed. This physician would regard your body as though it were a kind of machine, but he would not be very good at his job if he supposed it was 'nothing but' a machine, that you were other than a complete person with emotions, ideas, opinions and a unique life history; and the physician would be considered highly eccentric who regarded a human body as 'nothing but' a pair of kidneys writ large. The physician is not a scientist, of course, but a technician informed by discoveries made by scientists who investigated the human body – by taking it to pieces.

The essential difference lies in an important distinction between a mechanical system and one that is adaptive. A mechanical system, or machine, operates according to fixed principles structured into it at the time of its manufacture. Any deviation from those principles results in the failure of the machine. An adaptive system may also operate according to certain principles structured into it, but it has at least some flexibility. Should a component fail it may be able to compensate and continue functioning, or should its circumstances change it may be able to tolerate the change and survive in the new conditions. Up to a point, a living organism is a 'machine' that reacts with its surroundings and is able to make appropriate responses to changes in them.

Scientists do, indeed, study phenomena by taking them apart and examining each part separately. They do so because, after all, there is no other practical way they can be studied. There is a further, more fundamental reason for the adoption of reductionism. Recalling the hierarchy of understanding and the logic of transitive explanations, it is generally wise to assume that phenomena are not explained at one level in ways that contradict the findings at lower levels. As Peter and Jean Medawar put it, 'the actual is most easily understood as a special case of the possible'.[10]

When people abandon the reductionist approach in favour of holism, they reduce substantially the amount of understanding that can be extracted from their research. On 26 September 1993, eight people emerged into the Arizona desert after spending two years inside 'Biosphere 2'. This structure is an attempt to create a self-sustaining simulation of a terrestrial environment capable of providing humans with food and breathable air over a prolonged period, while disposing safely of their metabolic wastes, the Earth itself being 'Biosphere 1'. The idea is sound, if very ambitious, for it could deepen our understanding of the requirements for life on our own planet and provide invaluable information for those whose task it will be to design and equip living facilities for humans colonizing other planets or planning to live for long periods in space stations.

The Biosphere 2 project encountered several serious difficulties, but was criticized mainly for the poor quality of its scientific research. At fault, critics said, were the holistic ideas of its sponsors, who were 'far too steeped in New Age philosophies, theatre and art, cooperative living, organic farming and living an idyllic existence in space or on Mars'.[11] One consequence of this holism was the refusal to divide the Biosphere into sections so as to provide controls against which deviations could be measured and variables manipulated. What was reported as a puzzling event turned out to have a simple explanation when a geochemist, Jeff Severinghaus, was brought in from outside the project to examine it. Briefly, there was a serious decline in the oxygen content of the air and a quantity of carbon dioxide seemed to vanish. It transpired that soil micro-organisms, fed vast quantities of manure and organic matter, used more oxygen than had been predicted in the oxidation of carbon as they decomposed this material. This removed oxygen from the air and released carbon dioxide, which acidified the 'ocean'. The missing carbon dioxide reacted with calcium hydroxide in the concrete from which the structure was built, forming calcium carbonate. Had engineers and soil microbiologists been consulted, the problem might have been predicted. A new team of 'Biospherians' has now entered the complex, and changes have been made. This time scientists will be allowed to enter, stay for a while and then leave. The project is being placed on a sounder scientific footing, by making it more 'reductionist'.

Holism sounds good, but in practice it usually achieves little, though some of its disadvantages are gradually being overcome. Large-scale phenomena, such as an ecosystem or the body of a complex animal, can be studied only in terms of their parts, because to observe one detail it is necessary to exclude those others which might exert an uncontrollable influence on the observations. If all the relevant details, or at least the most important ones, can be understood individually, however, and testable assumptions can be made about their mutual interactions, then it may be possible to examine the whole system together. The result is a dynamic mathematical model, or representation, of the entire system. Components of the system can be allowed to perform a simulation of their ordinary functions, which include their responses to variations among the other components. If the results correspond to the phenomena observed in the real world, it can be assumed the model represents that aspect of the real world fairly accurately. Then changes can be introduced to determine the likely consequences. But the mathematics involved is formidable, partly for its complexity but mainly for the large number of individual calculations that are required to register even the smallest development. Sophisticated modelling of this order became practicable only with the introduction of fast, powerful computers. Even now, the simulation of very large-scale systems, such as the global climate, is limited by the computing power available (see chapter 17).

Insofar as computer modelling allows scientists to study complex phenomena as a whole, the technique may be described as holistic, but it depends on an extreme reductionism. If the model is to reflect accurately what it represents, it must be constructed as an assembly of its parts, each of which must be known separately. There are models, for example, that can predict reliably the movement and eventual fate of liquid effluents discharged into tidal water. But, to do so, every detail of the shores and beds, the flow channels and rates of flow of rivers and the tide, for all states of the tide, and the composition of the effluent and its reactivity and solubility in fresh, brackish and salt water must be measured. The purpose of such a model is to determine whether or not a particular discharge will harm the aquatic organisms and those, like birds, that feed on them. This requires measurements of the sensitivity to change of the aquatic community in general, determined by the individual sensitivities of its members and their relationships. In other words, the model is as reliable as the measurements on which it is based and its holistic results arise, and can only arise, from reductionist studies.

Hatred of reductionism dates from the early years of the last century and is based on a fear that the analysis of a phenomenon in some way diminishes or even destroys it. Willam Blake (1757–1827), for example, wrote in 'Mock on, Mock on Voltaire, Rousseau':

> The Atoms of Democritus
> And Newton's Particles of light
> Are sands upon the Red sea shore,
> Where Israel's tents do shine so bright.

Also referring to Newton and his experiments with a prism, which separated light into its constituent colours, John Keats (1795–1821) wrote in *Lamia*:

> There was an awful rainbow once in heaven:
> We know her woof, her texture; she is given
> In the dull catalogue of common things.
> Philosophy will clip an Angel's wings,
> Conquer all mysteries by rule and line,
> Empty the haunted air, and gnomed mine—
> Unweave a rainbow, as it erewhile made
> The tender-person'd Lamia melt into a shade.[12]

No scientist denies that the reductionist approach has its pitfalls and limitations. These are well understood. Should any scientist claim that any living organism is 'nothing more' than chemical reactions, or the spinning of electrons, then he or she is either joking or being silly. Scientists are but human, after all, and, like the rest of us, sometimes they say silly things.

Properly understood, with all the necessary intellectual safeguards in place, reductionism is perhaps the most powerful tool we have for understanding the, so to speak, weft and texture of ourselves and the universe that we inhabit. It is nothing to be afraid of and we should not be cowed by the modern holists who are trying to continue what was little more than a nineteenth-century Romantic literary propaganda campaign. Nor should we be seduced by them into supposing that it is possible to study the world, or a flower, simply by gazing reflectively at it. That may be enough for the purposes of the painter or poet, but it will reveal nothing of the structure, function and importance of the flower. The dissected description by the scientist enhances rather than detracts from both its beauty and our sense of wonder at its complexity and subtlety.

Philosophers and Physiocrats

Two centuries ago, when the Enlightenment was at its height and observations of the behaviour of natural phenomena were being organized into general principles, or 'laws', it must have seemed obvious that so productive a method could be applied much more widely. Scientific progress was rapid in an age when an educated person might well have been able to comprehend the latest developments, arguments, and discoveries across the full range of disciplines. Advances were being made in anatomy, entomology, hydraulics, astronomy, magnetism (including the geomagnetic field) and electricity, as well as in printing, manufacturing, engineering and construction, the 'practical arts', and especially mathematics, including probability and the beginning of statistics. For those who found difficulty with the mathematics and formal arguments, there were popularizations, such as *Newtonianismo per le Dame* ('Newtonianism for the Ladies') by Francesco Algarotti, published in 1735, as well as translations of works written originally in Latin.

Those engaged in this great endeavour were called philosophers. The word 'science' existed, but was associated more with the acquisition and application of practical skills than with expanding knowledge. 'Scientist' did not come into use until around 1840. The philosophers moved freely among disciplines not yet delimited by boundaries and they soon began to relate recent discoveries to human affairs. Societies, for example, were said to be conditioned by the geographical terrain and climate in which they developed. Voltaire proposed a link between climate and temperament in 1734, in his *Traité de métaphysique*, and also suggested that terrain determines whether the inhabitants of a region are placid or belligerent. Charles Louis de Secondat Montesquieu (1689–1755) suggested, in his two-volume *De l'Esprit des lois, ou du rapport que les lois doivent avoir avec la constitution de chaque gouvernement, les moeurs, le climat, la religion, le commerce, etc.* published in 1748, maintained that the Protestantism of northern Europe was due to climatic and political causes.[13]

Amid this widespread application to human society of the principles being elucidated in respect of natural phenomena, an entirely new discipline began to emerge. Economics, formerly considered only as a branch of politics, became established in its own right. In Britain, the first chair in political economy was created in 1805 at Haileybury College, owned by the East India Company, and its first professor was Thomas Robert Malthus (see chapter 15). However, it was in France that the movement began.

In 1758, François Quesnay (1694–1774), court physician to Madame de Pompadour and Louis XV, published his *Tableau économique* in which he related industry to agriculture and proposed that only agriculture added to the wealth of a nation. In fact, he claimed to have discovered laws describing economic behaviour analogous to those describing natural phenomena and his ideas attracted supporters who formed the Physiocratic School. They maintained that agriculture is based on the production of something from nothing through the divine miracle whereby the land produces a surplus of food and raw materials for our use, together with the seeds for its own perpetuation. This, they said, forms the basis of all wealth, industry merely rearranging the primary produce, and the wealth thus produced circulates according to a natural harmony. If, through our reason, we can discover this harmony and bring our institutions into alignment with it, they said, society will be harmonious and living standards will improve in perpetuity. All of this the Physiocrats summed up in their phrase 'laissez faire, laissez passer', which enjoins governments to remove all burdensome taxes and regulations, and yield to the natural harmony. Developed further by Adam Smith, who removed the excessive agrarian bias in his *Inquiry into the Nature and Causes of the Wealth of Nations*, published in 1776, physiocratic ideas have provided the basis for conservative economic theories to the present day. They were supported by many leading economists, including David Ricardo (1772–1823) and John Stuart Mill (1806–73).

This story exposes the dangers inherent in all attempts to apply scientific concepts in areas where they are inappropriate. Such attempts require parallels to be drawn where none exist and the economic, social or political interpretation turns all too rapidly into a dogma derived from concepts that were misunderstood or incomplete. Call it a 'divine miracle' if you will, but the land does not produce something from nothing. The growth of plants is a cyclical process in which nutrients, some of them mineral compounds, are processed and must be replenished if crop yields are not to diminish. Industrial manufacture is similarly based on the processing of minerals taken from the ground. If there is a law, or general principle, describing economic systems it is not to be found in the biochemistry of plant growth or the physics and chemistry of manufacturing. Yet major economic theories continue to base themselves on a 200-year-old scientific fallacy.

Social Darwinism

'Survival of the fittest', the phrase that is often used to summarize the Darwinian theory of evolution, was coined not by Charles Darwin, but by Herbert Spencer (1820–1903) and it marked the next attempt to apply scientific ideas to social affairs.[14] Years before Darwin published his theory, Spencer had reached somewhat similar conclusions independently. In his scheme, evolution involves an increasing complexity, for example, from the single-celled organism within which there is little specialization of function, to the multi-celled plants and animals in which there is considerable differentiation into specialized parts. The concept is hierarchical and has two interesting implications: it implies progress, from simple to complex, from homogeneous to heterogeneous; and parallels can be drawn between biological evolution and the development of commercial or industrial undertakings from the small workshop or store to the vast factory or chain of department stores.[15]

Spencer was also influenced by the Malthusian theory of population, the physiocratic theory of economics and the principle of the conservation of energy. This was the first law of thermodynamics, which was first described in detail, in 1847, by the German physiologist and physicist Hermann Ludwig Ferdinand von Helmholtz (1821–94). Darwin and Alfred Russel Wallace proposed that random but heritable variations among members of a species equip some better than others to find mates and reproduce success-fully under the conditions in which they all live, thus perpetuating in offspring those traits that proved advantageous. This natural selection from the pool of variation is the driving force of evolution. Accepting this, Spencer added to it the idea that evolution is progressive, that, by and large, later forms are superior to earlier ones. Accordingly, the evolutionary process is to be applauded and encouraged. It can be encouraged by intensifying selection pressure, which may achieve little by way of short-term biological improvement, for evolution is understood as a very gradual process. However, much more rapid improvement, said Spencer, could be anticipated by applying the theory to social and economic affairs. If competition is increased greatly, the survivors will be the strongest and, within a few generations, the population as a whole will be stronger, in an economic rather than biological sense. Moreover, the improvement will be manifest not merely in increasing complexity, but in the better adaptation of individuals, companies and institutions to the economic environment in which they find themselves.

Clearly, the economic theory best equipped to bring about this desirable condition is the one proposed by the Physiocrats and their descendants, who believed in the existence of a natural law governing economic behaviour. So, either Spencer and his followers had discovered this law or, at the very least, they had found a means of observing it and so achieving the natural harmony it promised. By a quirk of history, the credit, if that is the appropriate term, went not to Spencer, but to Darwin. The result is what came to be known as Social Darwinism and, like the physiocratic ideas on which it is partly based, its influence endures to the present day. All talk of 'welfare scroungers', 'encouraging people to stand on their own feet' and 'penalizing the work-shy' might have been taken directly from Social Darwinist texts. In 1914, William Graham Sumner of Yale University wrote that: 'The millionaires are a product of natural selection. . . . They may fairly be regarded as the naturally selected agents of society for certain work.'[16]

Scientifically, the Social Darwinist theory is deeply flawed. Its first error lies in its equation of evolution with progress, the idea that later forms are better than earlier ones. This is a value judgement, for what do we mean by 'better'? If we mean 'better at surviving' we are being tautologous, since all this tells us is that some survive because they are better at surviving. How else, though, are we to define 'better'? Biologists are in no danger of falling into this trap, because they do not allow the initial equation. There really is no meaningful sense in which, say, a modern horse can be described as better than a hadrosaur, which lived in the Cretaceous Period. Both were pretty well adapted to their conditions, but along with the other dinosaurs, the hadrosaurs were unlucky and so far luck has not run out for horses. Is a human biologically 'better' than a single-celled alga? The alga has been around much longer, so it might be more sensible to call it the more successful. It might be, but, of course, the whole idea is a nonsense. In a lecture on 'Violence and machismo', psychologist Richard Ryder suggested that: 'The whole doctrine of the market economy is . . . based upon a macho-driven misunderstanding of Darwinism – the survival of the strongest rather than the fittest. Strong and fit are not the same.'[17]

Social Darwinism also falls at later hurdles. Biological evolution involves the change from one species to another, but Social Darwinism is concerned only with humans and their institutions; it does not propose that humans evolve into post-humans. Indeed, in its scheme of things, that might require humans to become extinct through inferiority, which seems to defeat the object. No valid parallel can be drawn between the biological and social theories.

The Social Darwinists err on a third count, of misinterpreting the evolutionary concept of 'competition'. Biologically, this is firmly restricted to inter- or intraspecific competition for such resources as may be required for breeding and raising young. Non-humans do not grow rich or expropriate more resources than they need. In any case, the Darwinian emphasis on competition, derived from nineteenth-century combative expansionism, was overdone. Among and within species, cooperation is probably commoner than competition and at least as significant.

The fallacy of eugenics

Social Darwinism was a popular political idea that was founded on a scientific fallacy. It was not the last, for apparent scientific endorsement of short cuts to self-improvement can usually command support. In a book published in 1883, Francis Galton proposed a scheme 'to further evolution, especially that of the human race'.[18] He had in mind the elimination of heritable defects by selective breeding, which he believed would result in a human stock greatly improved in physique and beauty. However, he claimed: 'It is with the innate moral and intellectual faculties that the book is chiefly concerned.' Galton called the enterprise 'eugenics'. He founded the English Eugenics Society in 1907, two years after Madison Grant, Henry H. Laughlin, Irving Fisher, Henry Fairfield Osborn and Henry Crampton had founded the American Eugenics Society. In the 1920s and 1930s eugenics laws passed in several European countries and in 27 US states permitted the sterilization of 'defective' persons. In a court case in the United States in 1927, the judge, Oliver Wendell Holmes, authorized the sterilization of a woman with the statement that 'three generations of imbeciles is enough'.[19] Konrad Lorenz, the animal behaviourist, advocated 'racial hygiene' for the 'elimination of morally inferior human beings'[20] and many other eminent scientists held similar views. Nazi enthusiasm for eugenics blunted its popularity for a time, but its superficial persuasiveness ensures its periodic reemergence from the obscurity it deserves.

Eugenics may be the one political idea masquerading as scientific that has attracted serious scientists. Like Spencer, Galton supposed that the taxonomic hierarchies into which biologists and palaeontologists place living or formerly living organisms have a moral significance, that those occupying higher hierarchical levels are 'better' than those at lower levels, and he defined 'better' in moral and intellectual terms. One of the founders of statistics, he devised a great variety of tests to measure the differences he believed important. Most of his tests fell from use long ago, but we have retained our love of testing. Stephen Jay Gould has explored in detail the misconceptions that have bedevilled the most popular of them all, intelligence testing.[21]

As you might expect, we generally class as 'desirable' those qualities by which we identify ourselves. The 'best' people, therefore, often turn out to be those who belong to the ethnic group and social class presently enjoying dominion over the rest. In the United States, the Immigration Restriction Act, 1924, imposed strict limits on entry to the country in the case of people bearing 'undesirable' characteristics. These were held to be prevalent in countries that lacked the good fortune to be located in northern Europe. The ideas underlying this legislation were also used to justify racial segregation.

Today, most people would consider this crude, but without necessarily rejecting the eugenicist thesis altogether. It is often argued, for example, that improvements in health care permit the survival of individuals with severe inherited disorders or disabilities and that this weakens the population at large. It can be explained, of course, that what matters in evolutionary terms is not heritable disabilities so much as the likelihood of their being passed on to future generations. Most of the more serious disabilities prevent the sufferer from reproducing, so the problem is less serious than it may seem. The fall-back position is to argue that the genetically disabled are being artificially favoured in the competition for scarce resources, to the disadvantage of, if you will, the advantaged. Expressed this way, the desperation behind the argument is apparent. It is ridiculous, for unless society had sufficient resources the health care would not be available in the first place.

Modern advances in genetics have led to proposals for, and fears of, a new eugenics programme. Since certain genetic traits can be defined as desirable and others as undesirable, if persons with desirable traits were encouraged to breed, while those with undesirable traits were discouraged from breeding, over a number of generations the population would enjoy the genetic improvement resulting from the elimination of deleterious genes.

With any luck, this apparently sophisticated version of the old idea may prove its final undoing. It begins with the old, sweeping assumption that 'desirable' and 'undesirable' are easily definable. This may be true for plant and livestock breeders seeking to produce flowers with new colours, cows that give more milk or chickens that do not go broody, but in humans the selection is more difficult. If we decided to breed for a particular hair colour, that might be feasible, but everyone would think it pointless. If we prefer to breed for intelligence, say, or a highly developed moral sense, we face a difficulty, because no one knows how far these traits are subject to genetic influence and everyone agrees they are not wholly determined genetically. Breeding for them, and the converse, breeding out such traits as criminality and stupidity, is impossible. To some extent Beethoven, Einstein and Adolf Hitler inherited the characteristics by which we remember them, but had they not grown up and lived as adults in the particular social environments they did, those characteristics might never have been expressed and we would never have heard of them.

Less ambitiously, people might suppose it feasible to eliminate genetic abnormalities that lead to clearly diagnosable illness. It is in this sense that 'genetic eugenics' is usually understood. It is very probable that within the next few years it will possible to identify many such genes and replace them with sound genes in individuals liable to develop those illnesses. Whether such manipulation will ever be practised raises ethical questions that society must decide. At present, gene therapy of this type is illegal in most countries. In any case, even if it were allowed treatment would be on an individual basis, it would not and could not eliminate those genes from the population at large. Furthermore, they could not be eliminated by any kind of breeding regime.

This is because each of us inherits two sets of genes, one from each of our parents. We possess our genes in double doses and in each case only one of the pair of genes is activated. A gene may be dominant or recessive, and if it is recessive it will have no effect on us whatever. A recessive deleterious gene will be expressed only in an individual who inherits two copies of it (is homozygous for it). Such an individual might be helped, but others, who carry only one copy (are heterozygous) will remain unharmed. Some 4,000 potentially deleterious genes have been identified so far in humans and we all carry some of them. Therefore, a breeding programme aimed at eliminating them would forbid anyone to breed. This is the nonsensical conclusion to which the most favourable interpretation of eugenics eventually leads.

Happily, there are few examples of such misapplications of ideas derived from distortions of scientific discoveries. But the fact that there are so few should not lull us into complacency. A new one could emerge at any time. I have in mind the construction of social theories based on quantum theory[22] or the parallels sometimes drawn between quantum theory and oriental religions.[23] They seem harmless, and probably they are, but the track record so far is not encouraging. Beware, then, grand propositions for personal, social, economic or political improvement whose proponents claim a scientific basis. Distrust non-scientists whose nostrums have been 'scientifically proved'. Disbelieve those who claim to have found short cuts to Utopia. We can improve, progress is possible, but the road is long and sometimes hard.

If we permit them to do so, scientists can contribute greatly to our efforts by increasing our understanding of all aspects of the universe, and so assist us in avoiding errors or attempting the impossible. But scientific explanations refer only to the phenomena they describe and on no account should they be generalized or taken out of context. We are not social insects, and the society of ants or bees supplies no model relevant to our relationship with one another. Societies do not evolve by natural selection in the way species do. Selective breeding produces attractive and productive plants, farm livestock and household pets, but humans cannot be bred selectively. Nor are we fundamental particles inhabiting a quantum world of apparent irrationality. Trust and believe what scientists say, but at all costs avoid reading into their statements meanings that are not there and were never intended by their authors.

Scepticism and truth

We live in an age of doubt and applaud those who question 'accepted truths' or 'eternal verities'. Scepticism is always healthy, but sometimes it can lead to paradoxes of which not everyone is aware. It is fashionable among anti-scientists, for example, to challenge the idea that scientists are engaged in a quest to reveal the truth by means of an objective investigation of natural phenomena. Their challenge turns on the words 'truth' and 'objective,' leading them to conclude that the scientific description of the universe is but one description among many and is neither more nor less valid, or likely to be true, than any other. The observations from which scientific statements are made, it is argued, are far from objective: scientists select the phenomena they observe and tailor their observations and theories to support pre-existing ideas. Thus the enterprise is entirely subjective, they say, and there are no objective standards by which the truth can be determined.

According to Václav Havel, playwright, philosopher and president of the Czech Republic, 'Modern thought – based on the premise that the world is objectively knowable, and that the knowledge so obtained can be absolutely generalized – has come to a final crisis. This era has created the first technical civilization, but it has reached the point beyond which the abyss begins'.[24] In other words, claims to objectivity are false and therefore subjectivity is the only mode available to us for comprehending the universe. Clearly, acceptance of subjectivity implies 'trusting our feelings' or our 'intuition' as more reliable guides to reality than rational thought and logical argument. In the words of another philosopher, Paul Karl Feyerabend, 'anything goes'.[25] Feyerabend maintains that 'normal science is a fairy tale' and that 'equal time should be given to competing avenues of knowledge, such as astrology, acupuncture, and witchcraft'.[26] He has also supported the teaching of 'creation science' in schools.

As human beings, obviously scientists think and feel subjectively. To this extent they are indeed subjective. This does not imply, however, either that no reality exists outside ourselves, objectively, or that the methods employed by scientists are incapable of revealing it and should be abandoned. Much of the supposed rivalry between subjectivity and objectivity arises from very old-fashioned ideas about the essential character of scientific investigation, in which anti-scientists believe they detect a weakness that can be exploited. Observing that scientific ideas change over time, they suggest that either the reality those ideas describe is also changing, and consequently is unpredictable and probably unknowable, or reality is unchanging but the ideas were false in the past and are likely to be false now; in both cases the scientific claim to objectivity is spurious. The issue centres on the concept of irrefutability.

Sometimes we hear people describe this or that fact as 'irrefutable' and we tend to associate 'irrefutable facts' with 'scientific facts'. This association marks the gate onto the road we have travelled to reach our present position, a gate first opened in modern times by Sir Karl Popper.[27] Disliking certain theories that claimed to be scientific, in particular Marxism, Popper found them to be irrefutable. In contrast, Einstein's theories could be used to make predictions that would turn out to be valid or invalid, so the theory itself could be refuted. This led him to propose that irrefutability is a vice, not a virtue, and a theory can be said to be scientific only if, at least in principle if not in practice, it is capable of refutation.

In reaching this conclusion, Popper demolished an old description of what it is that scientists do. This is the idea, which may have originated with Francis Bacon, that scientists make observations of particular phenomena and accumulate their results. Eventually, someone discerns a pattern among the observations and this leads to the formulation of a general explanation, or 'theory', that encompasses them all. Instances are then sought to confirm the theory. The method is called 'inductionism', and in Popper's view it is entirely wrong.

Popper was addressing an old and well-known difficulty concerning the way generalized descriptions, or 'laws', are reached. If we observe that event A is followed by event B, our observation relates to only one instance and we cannot generalize from it. No matter how many times we repeat the observation and B continues to follow after A, we can never be certain that some future observation will not dissociate the two. Bryan Magee has illustrated this, and Popper's solution to it, with the example of swans.[28] No matter how many white swans we see, we cannot justify the general statement 'all swans are white', because somewhere in the world there may be a swan that is not white, or at some time in the past such a swan may have existed or may exist in the future. If we see a black swan, however, we can refute at once the statement 'all swans are white'. There is, therefore, a class of statements that can never be finally verified, but can be refuted. These are the statements Popper calls 'scientific'. Many scientific 'laws' associate causes and effects to which no exception has ever been found. This makes predictions based on them very reliable, but in principle they could be refuted. Inductionism, therefore, is an illogical procedure and cannot be regarded as the underlying basis for scientific discovery.

There is a danger, however, that should such an exception be found it might be dismissed or reclassified in order to protect the general statement. The black swan, for example, might be classified as something other than a swan, or its colour might be attributed to some extraneous cause, such as a mutation peculiar to itself, which failed to undermine the 'law'. This would not be inconsistent, and therefore refutation cannot always be finally conclusive, but neither can we allow observations to be reinterpreted endlessly. To avoid this dilemma, Popper proposed only that scientists do not go out of their way to dismiss anomalous data and that they formulate theories in ways that allow for possible refutation.

137

Inductionism can also be rejected as a description of what actually happens. Theories do not suddenly appear out of the data, as it were. Scientists begin their observations with theories already in place. They must do, because they must have had some reason for selecting in the first instance the phenomena they consider worth of study. These must interest them or offer the possibility of explanation, and ideas of this sort are themselves theories. Asked where such theories come from, Popper replies that they spring from still earlier theories, in a regression that takes the individual back to early childhood, and childish ideas of the way the world is, and culturally back to myths, which in some sense are also theories. The scientific task, therefore, consists in the critical examination of existing theories and, where appropriate, their replacement by better theories.

If theories can neither be finally verified nor conclusively refuted, we are at once liberated and constrained. We are liberated because the way is open for new theories to supersede older ones and for partial theories to remain useful even though they appear to contradict one another. We are constrained because scientific theories can never be described as true. Scientists may strive to attain truth, but it is an iterative process that may never reach its declared goal. Whether or not such a thing as 'Truth' exists, in an absolute sense with a capital T, is a question for philosophers rather than scientists. The absence of any kind of final, categorical statement of Truth does not leave us bereft of guidance, because scientists deal much less with Truth than with probability. A statement, or 'natural law', may or may not be True in an absolute sense, but there is a certain probability of its being true. It is possible, for example, that all the atoms that make up the air in the room where I am now sitting may suddenly congregate in one corner, leaving me, rather literally, breathless. It is possible, but so very improbable that I need take no account of it. It is possible that if I throw a ball into the air it will continue to rise for ever, but no instance of this has ever been recorded in the whole of human history. It is a highly improbable event and everyday familiarity with the local effects of gravity will, almost certainly, serve me for the remainder of my life just as well as it has so far served me and my ancestors back to the dawn of time.

Theories, paradigms and postmodernism

'Theory' is also a word that causes some confusion. In ordinary conversation someone might say, 'I have a theory that . . .' or 'what you say is all very well in theory, but . . .' and mean by 'theory' a purely speculative idea that explains something independently of the facts to which it refers. This kind of 'theory' is what a scientist would call a 'hypothesis': a general idea advanced as a possible explanation for certain phenomena, but with no claim to truth; it is an idea offered up for testing against observations. A scientific 'theory' is a statement linking a number of instances of an observed phenomenon into a generalized explanation from which predictions can be made. Biological evolution, for example, is a phenomenon that has been observed in many species. The Darwinian theory of evolution gathers together those observations and supplies a general explanation of the process giving rise to them whereby the presence of the factors driving the process can be predicted to result in the formation of new species.

Scientists do not go around observing just anything. They select what they observe for its relevance to the theory that it will fit into. The theory is essential, but this raises further difficulties. Theories are abstract constructions and subjective. If observations are so closely related to theories that, in some sense, they are produced by them, then the observations too acquire something of the subjective character of theories. This is the argument used by critics to challenge ideas of objectivity and, from it to conclude that, if scientific observations are subjective, they have no greater claim to authority than any other subjective construct. Even worse, this subjectivity undermines even the Popperian idea of refutability, for it is clearly impossible for one theory to refute another.[29] Thomas Kuhn, in The Structure of Scientific Revolutions, pursued the problem further. He concluded that scientists compile theories into which they fit the data that they collect. Everything proceeds peacefully within this paradigm until the paradigm, or 'world view', is suddenly and dramatically overturned and a new one replaces it. Scientists are not pursuing truth so much as conforming to contemporary intellectual fashion. Their theories are ephemeral.

With doubt cast on the concepts of truth, objectivity and refutability, the entire intellectual rug seems to have been pulled from beneath the scientific enterprise. 'Ice floats on the surface of water because the density of water is at a maximum at $4°C$' and 'The Moon is made of cheese' become statements of equal validity. Neither can be objective, because the possibility of objectivity is denied, we cannot say whether or not either or both of them is true, because we no longer accept the concept of objective truth, and both are equally susceptible to inconclusive refutation. Everything has become relative and we are left wondering just what it is that scientists are doing.

It is the conjunction of these ideas that has fostered the popular anti-scientific view, that scientists are propounding an ideology that should be judged in the same way as any other ideology. We are free to accept or reject any or all of it and to prefer any other ideology that may have greater appeal. According to this argument, anti-rational relativism, often known as 'postmodernism', is also an ideology, and it is here that we begin to discover certain paradoxes. The most obvious of these concerns the rejection of truth. This is tantamount to saying, no proposition is true. But is that proposition itself true? If it is true it is false and that is the paradox. Put another way, if there is nothing of which we can be certain, then neither can we be certain of the statement 'there is nothing of which we can be certain'.

If scientists are engaged in supporting contemporary paradigms, by whose authority are these paradigms sanctioned unless by scientists themselves who find in the world views they enshrine coherent and internally consistent explanations of phenomena? Do the revolutions by which paradigms are overturned amount to mere changes in fashion, or do they result from new information, new discoveries that cast phenomena in a different light permitting the construction of a more comprehensive world view? Anti-rationalists also support their own world view, or paradigm, but its sanction derives only from a rejection of the supposed basis of all scientific statements. Anti-rationalism can be true only if scientific paradigms are false, but since it rejects the attribution of truth to either, its truth lies in its denial of truth. This is another paradox. The anti-scientific edifice is a deal less robust than it may appear.

It is not only among scientists that the idea of objective truth is important. Every aspect of ordinary life requires us to observe and to construct a rational interpretation of our observations that allows us to make predictions. How far the model that each of us builds of the world we live in corresponds to phenomena outside ourselves determines our ability to live successfully in the world. The model must provide at least a measure of its truthfulness. For example, I do not walk off the edge of a cliff, because my world model predicts that doing so would probably lead to my death. Each of us has a set of political opinions. According to these, we urge politicians to pursue this or that policy because our world models allow us to predict the possible outcomes of different courses of action. We may be wrong in our understanding of the implications of alternative policies, but we accept the existence of a real world, containing real people, who might be affected in the way we suppose if our understanding of the policies we were considering were correct. Our idea that there is a real world is surely true. It is difficult to see how those who deny the existence of any kind of objective truth are able to function at all, for they must attribute equal value

140

to any and every opinion while denying the validity of their own observations. Socially as well as intellectually, this can lead only to chaos.

As an ideology, anti-rationalism is inconsistent and its practical application impossible. Since its proponents appear to function normally, we are bound to assume either that they genuinely believe in the existence of a coherent, to some extent predictable reality outside themselves, or that they behave as though this is what they believe. Not only is their position inconsistent, therefore, but holding it is hypocritical.

There is another direction from which scientists have been attacked. Like everyone else, scientists make statements of different kinds. Sometimes they describe what they or others have discovered and sometimes they speculate. Most speculations are simply ideas, or perhaps hypotheses, but on occasion they can be grander and more abstract. Confusing these different types of statement gives rise to an entirely new area for criticism.

The anthropic principle

Possibly the grandest of all contemporary concepts is the 'anthropic principle', an idea developed by some cosmologists. There are two versions of it, 'weak' and 'strong'. According to the weak anthropic principle, certain physical relationships happen to be such as to permit the existence of matter and hence that of a universe which includes intelligent observers. The charge on a proton, for example, is precisely equal to that on an electron, but there is no particular reason why it should be so. However, if it were otherwise stars would not be luminous and atoms might not have formed at all. This is one of a number of such apparent coincidences all of which lead to the conclusion that the universe is the way we see it because we exist to observe it. Stephen Hawking likens it to the rich person living in a wealthy neighbourhood who sees no poverty.[30]

The strong version goes much further. This holds that the universe we see is only one of many possible universes or regions of a single universe, but it is the only one configured in such a way as to allow the emergence of intelligent beings. The universe is the way we see it because 'if it had been different, we would not be here'.[31] Some have taken this argument further, suggesting that the configuration of the universe in which intelligent life emerges makes its emergence inevitable; that it is the existence of intelligent observers that confers a tangible reality on the universe — the universe requires observers. Consequently, once those observers exist and commence

141

the compilation of information about the universe, they and their enterprise cannot die out.[32] In *Science as Salvation*, Mary Midgley bases an entire critique on this cosmological speculation. Her principal complaint is that it is essentially metaphysical and, in her view, bad metaphysics. She also points out that in its strong form it is teleological, but maintains that all scientific theories are to some extent teleological and justifiably so.

Her case is partly valid, but it cannot sustain any kind of critique of science because her target is itself arguably unscientific. Indeed, some scientists maintain that theoretical cosmology, being based on little or no direct observation, is metaphysical, as Midgley argues, but that this disqualifies it from consideration as a scientific discipline and statements deriving from it are not scientific statements. Even among those who would not go so far, the anthropic principle attracts only limited interest. In its weak form it is tautological, for it states that our existence proves that the universe is such as to contain ourselves. In its strong form it is teleological, stating that our existence is the purpose for which the universe exists.

I referred in chapter 10 to the gravity with which scientists regard the charge of teleology. Midgley considers teleology essential: '. . . the idea of dropping it altogether may not be much more practical than that of stopping breathing'.[33] This is true, but only in a trivial sense. Obviously, the scientist who plans an experiment does so with a purpose in mind, so the work itself is to that degree teleological with regard to the individual experimenter. This is true for any planned activity undertaken by anyone: indeed, we often describe such activity as 'purposeful'. That is not at all the same thing as ascribing a purpose to external phenomena, which is the meaning scientists attribute to the term and which they reject because it leads to absurd conclusions. These were most succinctly exposed by Mark Twain in a way pertinent to the strong anthropic principle:

> Man has been here 32,000 years. That it took a hundred million years to prepare the world for him is proof that that is what it was done for. I suppose it is. I dunno. If the Eiffel Tower were now representing the world's age, the skin of paint on the pinnacle-knob at its summit would represent man's share of that age; and anybody would perceive that that skin was what the tower was built for. I reckon they would, I dunno.[34]

Doubt can be constructive if it permits the serious reconsideration of ideas and the possibility of advance beyond them. The postmodernist, anti-rationalist critique put forward by anti-scientists is not constructive. It is frivolous, inconsistent and 'in the consequent intellectual chaos there results a thick fog of confusion which completely obscures every possible route to further discoveries, and in this way the quest to enlarge knowledge is effectively paralysed'.[35] We can break free from this obscurantism by exposing its serious intellectual weaknesses while at the same time reasserting the validity of the basis of the scientific endeavour. This is entirely reasonable. It requires us to accept the reality of phenomena outside ourselves with which we interact, and that surely can trouble no one. These external phenomena are coherent and to some extent predictable. Increasing our understanding of them will enrich us culturally while helping us in our everyday lives.

Scientists have developed the most effective means available to us for achieving this aim. We should recognize that the ultimate goal is the attainment of truth, though this may never be fully realized, but the discoveries made along the way can be accepted as partially and provisionally true; they are more or less reliable guides. The scientific enterprise is therefore legitimate and deserving of our wholehearted support.

13

Scientists as Citizens

We bring order to the world about us by classifying things. Were we unable to consider things as members of classes, or groups, we would face the impossible task of regarding every object, happening, or idea as unique. The effort would overwhelm us and the world would make no sense at all.

Classification is necessary, but we are inclined to apply it over-zealously and nowhere is this more obvious than in our 'pigeon-holing' of people according to their occupations, the more remote the occupation from our own experience the more stereotyped our idea of those who practise it. Stereotyping leads us to two assumptions: that all those practising a particular occupation are similar in many important respects; and that their 'occupational character' pursues them outside working hours, so that to be, say, a widget-maker you must be a special kind of person, a 'widget-making sort of person' the whole of the time because that is the kind of person you are. There may be some truth in this, but people are not so uncomplicated. As well as being widget-makers or whatever we may be, all of us are also citizens with interests and opinions ranging over fields far removed from widgets.

Scientists are no different in this respect from anyone else. They have hobbies, take holidays, play with their children and form opinions about the issues of their day. Those opinions are no more or less deserving of our consideration than opinions held by you or me, except when they refer to the discipline in which the scientist works, because then they are informed by specialist knowledge not possessed by outsiders.

If you need legal advice, you go to a lawyer. But the law firms that are best qualified to discourage offensive behaviour by your neighbour may be less helpful in defending you against an indictment for murder. If you need scientific advice, the same applies. A molecular biologist may have interesting ideas about astrophysics, but if you are planning a visit to Saturn this may not be the best person to help you calculate your launch dates and orbits. Such is the mythology with which we surround scientists, however, that our stereotyping of them is often extreme. We fail to distinguish between the specialist whose statements are based on deep understanding and close familiarity with a topic and the individual whom we entice into less familiar territory. We seem to assume that anyone wearing a white coat and bearing the title 'doctor' or 'professor' must be a true polymath.

Scientists themselves are not free from blame. Many are intrigued by age-old puzzles and perhaps freer with their views on them than wisdom might dictate. One of the biggest puzzles, of course, concerns the origin of life on Earth. It turns on the question of how nucleic acids and proteins might have formed so as to produce self-replicating molecules and hence living cells, given that nucleic acids encode proteins, but can themselves form only in the presence of enzymes, which are proteins. There are several alternative theories, but they all assume that the process originated on Earth. It may have occurred in shallow water, on the surfaces of clay crystals or near hot volcanic vents, according to which theory you support, but once the chemistry began, evolutionary processes led to the diversity of organisms with which we are familiar. Questions remain over the length of time needed for very large, very complex molecules to form by pure chance. For biologists, the solution must lie in a mechanism such as the trapping of amino acid molecules[1] in clefts in strands of ribonucleic acid (RNA),[2] and, in the following evolutionary steps, the rate of mutation must be limited.[3] There must be an explanation, for the obvious reason that life exists.

Alternatively, some scientists maintain that life did not originate on Earth at all, but was introduced from elsewhere. They are driven to this conclusion by what they regard as the statistical implausibility of such a high degree of organization; that RNA molecules, comprising a minimum of several thousand smaller units, could have occurred by chance. Marcel J.E. Golay, for example, once illustrated this using the metaphor of a self-replicating robot with bins of components in which it must search at random for the items it needs.[4] Golay is an engineer. Sir Fred Hoyle and his colleague N. Wickramasinghe, who are astronomers, have suggested that the necessary complex molecules reached Earth from space, citing as evidence the known existence of organic compounds in comets and other extraterrestrial bodies. Their theory closely resembles that of 'panspermia', proposed in 1908 by the Swedish chemist Svante August Arrhenius (1859–1927).

These are instances of attempted contributions to biology from scientists who are not themselves biologists. Evolutionary biologists regard the matter as intriguing and mysterious, since no one was present to observe what actually happened and so theories are bound to be conjectural, but they have no great difficulty in accepting the idea of the emergence of complexity from

145

simplicity, which is explained adequately by natural selection. Although chemical alterations, or mutations, occur randomly, increasing complexity is not a random process: if it provides the molecule or organism with an advantage, natural selection preserves it. It is not at all like engineering, because each advantageous modification provides the unit to which the next modification occurs and in this way complex organisms can be shown to develop very rapidly indeed. Biologists also point out that locating the origin of life elsewhere than Earth does not solve the problem. We are still not told how the organization happened.

The overall effect is to create an apparent muddle where specialists in the field perceive only a mystery to be unravelled, a difficulty in practice but not in principle. Properly understood, natural selection provides an entirely convincing mechanism. Interventions from non-specialists, especially those who are eminent in their own fields, merely confuse non-scientists and unintentionally provide comfort for those who maintain the real alternative to a terrestrial origin and evolution is not some kind of panspermia, but the creation by God described in the Book of Genesis. On the other hand, as Francis Hitching has pointed out, there are so many possible rational explanations for the origin of life on Earth that the creation scientists may not improve their case by inviting too much scrutiny.[5]

Perhaps it is our impatience that leads us astray, having grown used to getting answers to all our questions and getting them quickly. Religious systems being complete descriptions, their spokespersons can supply answers to all questions, but scientists find it more difficult. Since their enterprise is incomplete, they lack a total description of reality that gives all the answers. They ask their questions one at a time and accept only those answers that can be securely located within the partial description they have constructed. It takes time, and there are many questions to which no answers have so far been found. Lacking the explanation we so clearly think we need, we are tempted to invent them, however implausible they may be. It is hard for us to accept the perfectly legitimate statement from those most qualified to make it, that 'we simply don't know'

Breaking the rules

Occasionally, puzzling things happen even to scientists working within the confines of their own disciplines. In 1988, a team led by Jacques Benvenist of the Université Paris-Sud reported that a certain type of antibody (known as anti-IgE) remained effective even when diluted in water to one part in 10^{12}

(i.e. 1 followed by 120 zeros).[6] At this high dilution it is probable that the liquid contains not a single molecule of the antibody: it is pure water. Dr Benveniste could not understand the results and neither could the editor of *Nature*, the journal that published his paper. There is a rule among experimental scientists that results must not be rejected simply because they inconveniently contradict theory, but those that do, even less spectacularly than these, should be checked very carefully. Before his paper was published, Dr Benveniste was asked to check his results and did so. His findings were so extraordinary that John Maddox, editor of *Nature*, Walter W. Stewart and James Randi visited his laboratory to conduct their own investigation. Dr Stewart is an authority on scientific misconduct and errors and inconsistencies in the scientific literature and James Randi is a magician with considerable experience of detecting deliberate fraud. Their findings led them to reject the Benveniste results, which they attributed to poor experimental design and lax laboratory practice.[7] However, Benveniste continued to defend them and the row rumbled on for some time.

Accepted at their face value, the results were interpreted as support for homeopathic medicine, in which highly diluted medicaments are used. 'In the light of our investigation,' Maddox wrote, 'we believe that such use amounts to misuse.' Homeopathy is one of the most popular of 'alternative' medicines. People, not least its own practitioners, want to believe in its efficacy and would welcome warmly any supportive evidence. That the results were received in this way is not surprising, but the only support they really provide is for the view that wishful thinking may lead the most dedicated workers astray.

The 'homeopathic' experiments attracted little publicity outside the scientific community. News of the work that did, which falls into a similar category, broke some months later. On 23 March 1989, at a press conference held by the University of Utah in Salt Lake City, Professors Stanley Pons of Utah and Martin Fleischmann of the University of Southampton, England, both eminent scientists with impressive achievements to their name, announced that they had achieved nuclear fusion at room temperature. If so, the implications would have been tremendous, for the world would have been provided with a means of generating virtually limitless amounts of electrical power so cheaply it might have been almost free to the consumer.

In this case, the scientists concerned began by breaking a cardinal scientific rule: they released their results to the press before they had published them in a scientific journal. This is not a question of elitism or even of etiquette. All reputable journals send reports they are considering for publication to outside referees in a process called 'peer review'. In this case the referee might have detected errors and inconsistencies at once and, because of the startling nature of the experimental results, they would certainly have asked for more evidence. In the event, the day after the Utah announcement laboratories all over the world began trying to repeat the experiments armed only with the account they had read in the newspapers.[8]

Pons and Fleischmann had passed an electric current through an electrode made of palladium, that had been immersed in a solution of salts dissolved in water enriched with deuterium.[9] Over several hours, deuterium atoms were absorbed into the crystal lattice of the palladium. This much was already well known. Eventually, so many of them were absorbed and they were held so closely together that it was claimed that some of them fused to form tritium, a variety of helium, with a release of energy. When the experiment was run for 100 hours it was said to have yielded four times the energy required to operate it. In fairness to Pons and Fleischmann, it must be said that they were not alone in their search for 'cold fusion'. The United States Department of Energy had already invested a considerable sum in researching the possibility. In the end the search proved fruitless and yet another dream of cheap, abundant energy faded. The controversy generated no more than a long list of research findings, letters to editors and even books.

. Most of those who sought to repeat the experiment failed, producing no results whatever. Scientists often produce results that prove unrepeatable, and the idea of cold fusion was dismissed as no more than a case of wishful thinking leading to lax experimentation. That remains a probable explanation, but one should bear in mind the many years of laboratory experience that these senior scientists brought to their work, and that work continues. Fleischmann and Pons have attracted funding and several other laboratories are also pursuing similar lines of investigation. Now that they have refined and enlarged the cells, it is said that the energy yield exceeds the energy required to operate them by a much greater margin than Fleischmann and Pons claimed in their original report. No one knows what is happening in the cells, except that the absence of nuclear reaction products makes it seem unlikely that any nuclear fusion is taking place.

The ozone layer

Wishful thinking can be negative as well as positive. So anxious are we to believe that our activities are damaging the world that we give credence to supporters of this thesis however qualified their gloomy prognostications may be. Take, for example, popular concern over the 'hole' in the ozone layer. Ozone, the allotrope of oxygen in which atoms combine in threes rather than the usual twos, is formed in the upper atmosphere by the energy of certain wavelengths of ultraviolet radiation from the Sun. The energy is absorbed by ordinary oxygen molecules, breaking the bond holding the two atoms together. The free atoms then combine temporarily with oxygen molecules to form triplets, which are ozone molecules, and these in turn break apart with further absorption of ultraviolet. This happens mainly at a very high altitude, but some mixing of stratospheric air carries air relatively enriched with ozone to a lower height, around 25 kilometres, where it forms the 'ozone layer'.

Since sunlight is essential for the formation of ozone, during the darkness of the polar winter it cannot form, so gradually the concentration of ozone decreases. Some years ago it was discovered that over Antarctica there is a further, rapid depletion of ozone just as the sunlight starts to return, in early spring. This depletion, as we all know, is associated with the presence of certain compounds containing chlorine, and especially chlorofluorocarbon compounds, or CFCs, and it occurs because of the peculiar conditions present at that time of year in the Antarctic stratosphere. Concern about the ozone layer is not new. It was expressed in the 1960s, when some people feared that exhausts from large fleets of supersonic transport aircraft flying at high altitudes would destroy ozone. Our fears derive from the belief that exposure to ultraviolet radiation at the wavelengths absorbed in the upper atmosphere is extremely harmful. Consequently, atmospheric scientists have warned of the biological damage we may expect from ozone depletion and have been supported by physicians (though the scientists most qualified to comment on the matter are biophysicists). In particular, they link this directly to the incidence of skin cancer and to damage to marine algae, the phytoplankton.

Certainly there has been an increase in the incidence of skin cancers. This is associated with the fashion for sunbathing, not the depletion of atmospheric ozone, which, despite the claims, does not lead inevitably to an increase in the amount of ultraviolet radiation penetrating to the surface. Many other compounds absorb ultraviolet radiation and, while there is an established link between ultraviolet exposure and non-malignant skin tumours, the link to malignant melanomas, which are much more serious, is less certain and certainly more complex. For one thing, they tend to appear on parts of the body which are not usually exposed to sunlight.

The effects of ultraviolet radiation on aquatic organisms are normally limited to the uppermost surface layer. However, in water that is unusually clear, it may penetrate below 20 metres and be linked to the bleaching of coral, through loss of some of their covering of symbiotic algae (zooxanthellae). Ordinarily, only zooxanthellae close to the surface are exposed to ultraviolet radiation and they protect themselves against it, but in very clear water zooxanthellae at greater depth may be harmed before they can produce the protective compound.[10] The supposition of severe biological damage seems to derive from experiments performed many years ago with bacteria, with the aim of reducing cross-infections in hospital wards. Under certain conditions, some bacteria were killed by ultraviolet, but under the conditions in which bacterial colonies occur naturally it is very unlikely that they would have been seriously harmed.[11]

The concentration of ozone in the air is very variable. It changes from season to season and even from day to day. If the ozone layer were to disappear completely over the temperate latitudes, people living there would be exposed to about the same amount of ultraviolet radiation as is ordinarily received in the tropics. If it is depleted to the extent that we fear, the increased exposure will be equivalent to what would be received by moving several hundred kilometres nearer the equator. If ultraviolet radiation is so hazardous, we should ask how people can survive moves into somewhat lower latitudes, say from Britain to France or from the north-eastern United States to Florida, and how living organisms manage to survive at all in equatorial regions. As we know, life there is tumultuously abundant.

Our fears have been much exaggerated, because we have been too incautious in questioning the scientific authenticity of extravagant claims. It is as though we wish to believe that we are destroying the very possibility of life. This attitude engenders a fear of the future, a sense of hopelessness in the face of problems almost too vast to comprehend and probably beyond our capacity to solve. As usual, the truth is less sensational and a good deal less alarming. Probably it would be wise to limit our interference with the chemistry of the upper atmosphere and steps have been taken to do so as a precautionary measure. These measures seem likely to succeed, but should they fail it does not mean life on our planet will come to an abrupt end. Indeed, the consequences might be quite difficult to measure. Meanwhile, the best way to avoid skin cancer is to give up the habit of sunbathing.

14

Pollution and the Fear of Numbers

When, and if, we reach the point where the rate of combustion exceeds the rate of photosynthesis, we shall not only have to worry about running out of oxygen at night and in winter, but the oxygen content of the atmosphere will actually decrease.[1]

Man has adopted the attitude that he is above nature, that he can control nature and force it to do his will, and that he can undo any of the damage which he does . . . He has taken the attitude that the solution to pollution is dilution and has poured solid particles into the air expecting the air currents to dilute them so they are no longer harmful to him.[2]

We have caused the extinction of many hundreds of animal species, ransacked the planet for fuel and now stand like brutish infants, gloating over this meteoric rise to ascendancy, on the brink of the final mass extinction and of effectively destroying this oasis of life in the solar system.

The first of these quotations is from an article published in 1970, the second was written in 1972, and the third is taken from a recent Greenpeace leaflet. Before proceeding further, it may be worthwhile to point out that the fear of depleting the air of oxygen through the burning of fuels was examined and dismissed as a 'nonproblem' at about the time of the first quotation: 'Calculations show that depletion of oxygen by burning all the recoverable fossil fuels in the world would reduce it only to 20.8 per cent' (from its present 20.946 per cent).[3]

The truth about pollution

Perhaps we should find reassurance in the fact that so little has changed over twenty years. Doomsters remain consistent in their hyperbole and their scant regard for evidence to support the sweeping statements with which they indict us all, and there is no sign of an end to this assault. In August 1993, for example, *The Guardian* newspaper reported that Greenpeace had found levels of dioxin inside a certain factory site that were far higher than those anyone else had been able to find. 'Dioxins cause cancer, genetic defects and a reduction in sperm count,' thundered the report, omitting to tell us the source of this disturbing information.

We first became alarmed about dioxin in July 1976, when an explosion in a herbicide factory at the village of Seveso, near Milan, released large amounts of it. The accident led to the Seveso Directive, from the European Commission, which limits the stocks of hazardous chemicals that can be stored on factory sites and obliges companies to notify the authorities, their own workers and local residents of the quantities and nature of the chemicals they hold. For a time, dioxin was routinely described as 'the most poisonous substance known', one of a long list of compounds honoured with that title.

There are several dioxins, but the commonest, and the one to which the name usually refers, is 2,3,7,8-tetrachlorodibenzo-p-dioxin, or TCDD for short. It is a by-product of the manufacture of the herbicide 2,4,5-T and is also produced when substances containing chlorine, such as polychlorinated biphenyls (PCBs), are incinerated at too low a temperature. Apart from the Seveso incident, dioxin existed as a contaminant in 'Agent Orange', the defoliant herbicide that was sprayed over large areas of Vietnam during the war there. In fact, dioxins are very widely distributed, especially in industrial urban areas.

People living near Seveso, who were exposed to very large doses, suffered from chloracne, a distressing skin complaint that disappeared after a time leaving no after-effects. Intensive research has suggested a possible link between dioxin and cancer, but the evidence is slim. All attempts to link the dioxin sprayed in Agent Orange with birth defects and illnesses suffered by Vietnam veterans have failed. In Missouri, children playing in dust contaminated by dioxin did become ill, but the dust also contained many other contaminants and it was impossible to distinguish the separate effects of each.[4] Dioxin is not a pleasant substance and certainly emissions of it should be prevented so far as possible, but it is hardly the life-threatening bogey portrayed by environmental extremists, and in no way can the dubious

Greenpeace data support the emotive assertion by Debbie Adams, their 'toxics campaigner', that 'the Government is sanctioning a chemical experiment with the people and environment of Derbyshire the unwitting victims'. This is irresponsible alarmism.

Dioxin, or rather its absence, has proved profitable for some. Fears that disposable nappies might contain minute traces of dioxin picked up in the bleaching of the absorbent paper lining led one leading manufacturer to boast of its 'green' nappies that contained no dioxin. Its linings were made from pulp bleached by a process that used no chlorine and produced no dioxin. Investigations showed the claim to be correct, but the effluent discharged into rivers from its 'green' bleaching process killed more fish than effluents from the chlorine process, because they contained more of other pollutants.[5] Adapting an old proverb, one is tempted to say 'more greenery less greenness'.

Our fear of cancer provides fertile soil for propagandists, especially since it can so easily be fortified by our ignorance of simple arithmetic. In 1993, for example, the British Government proposed to its European Community partners that limits for pesticide residues be relaxed to those proposed by the World Health Organization. The existing EC limit for some pesticide products is set at 0.01 micrograms per litre of water. This may sound reasonable enough if you fail to appreciate just what that number means: one part in 100 billion (100,000 million). Not only is this a concentration so small as to be barely imaginable, it is also about half the minimum level that can be detected by the most sensitive analytical techniques. Indeed, it is known technically as 'surrogate zero'. To ingest one gram of pesticide at this concentration you would have to drink 100 million litres (22 million gallons) of water.

In the United States, the safety of pesticide residues in food is determined by the Delaney Clause, an amendment to the Food, Drug and Cosmetic Act introduced some thirty-five years ago. The Clause forbids interstate commerce in any food containing detectable amounts of any substance not present naturally that can be shown to cause cancer in experimental animals. This piece of legislation is likely to be revised, probably in 1994, but meanwhile it remains a nonsense. As long ago as 1969, the report of an official commission on pesticides said of it: 'If this clause were to be enforced for pesticide residues, it would outlaw most food of animal origin including all meat, all dairy products (milk, butter, ice cream, cheese, etc.), eggs, fowl, and fish . . . Removal of these foods would present a far worse hazard to health than uncertain carcinogenic risk of these trace amounts.'[6] The difficulty arises because the Clause stipulates no levels, and in sufficient amounts any substance can cause cancer in experimental animals. Interpreted literally, therefore, the Clause would ban all additives and preservatives as well as pesticides, leaving consumers either deprived of food or exposed to the (entirely natural) fungal and bacterial toxins most of the additives are designed to eliminate.

Confusion also arises from a misunderstanding of the most basic rule of toxicology: the toxicity of any substance depends entirely on the dose. Substances that are harmful in large amounts are quite innocuous in small doses. This is true for all poisons. Of course, the size of the dose that can cause harm varies widely from one substance to another and individual susceptibility also varies, but there is no substance so poisonous that a single molecule of it is likely to prove dangerous.

Common sense remains the rarest of commodities, however. Bruce Ames is possibly the most experienced cancer researcher in the world and the inventor of the widely used Ames test for identifying carcinogens. He is Professor of Biochemistry and Molecular Biology and Director of the Environmental Health Sciences Center at the University of California at Berkeley, a member of the National Academy of Sciences and the Royal Swedish Academy of Sciences, and winner of several prizes for his research into cancer. Professor Ames was invited to contribute an article for publication in the May 1992 edition of the magazine of the Sierra Club, a leading American conservation and environmentalist society. In it, he attacked the widespread misunderstanding and misuse of scientific concepts and data being employed by environmentalists. 'Pollution is being blamed for global warming and ozone depletion, pesticides for cancer,' he wrote. 'Yet these and many other environmental causes are based on weak or bad science. The reality is that the future of the planet has never been brighter.'[7] The Sierra Club rejected his article, but assured him this was not for its lack of 'ideological purity'. As Dr Ames discovered, exposing the purity as somewhat less than pure can be difficult. Few people want to have their prejudices contradicted and if the situation is even remotely as serious as it is described, then prudence dictates that we pay close attention to the message. The effect is insidious, for the truth of the message is seldom questioned.

Exaggerated fears

In most democratic countries, teachers are forbidden to use lessons for the political indoctrination of their pupils, but environmentalist literature crosses the barrier usually unobserved. 'Your parents and grandparents have made a mess of looking after the earth,' wrote Jonathon Porritt, former director of the British branch of Friends of the Earth and then leader of the Green Party, in *Captain Eco and the Fate of the Earth*. 'They may deny it, but they're little more than vandals. And they're stealing your future from under your noses.' This is but one of many books and television programmes directed at children. It is political propaganda which, so far from even pretending impartiality, simply ignores the possibility of any point of view but that of its author.[8] Its aim is to inculcate in its young readers attitudes that will pressurize parents and bring Green political parties to power. Indeed, it encourages young people to despise members of their parents' generation and, under Green rule, presumably to spy on them and report their misdemeanours to the ecological authorities. You may or may not think this desirable, but to mask the political message with a pseudo-scientific gloss is dishonest.

Nor does the message always make sense. The urge to recycle materials, for example, has been promoted to the status of dogma and it has caused environmental problems at least as serious as those it aims to curtail. In Germany, where householders are required to separate domestic wastes before collection, by the summer of 1992 the Dual System Germany disposal scheme had accumulated around 400,000 tonnes of plastic waste, compared with the less than 200,000 tonnes annual capacity of the recycling plants, and a DM500 million debt. Plastics are so cheap to make that it is doubtful whether recycling them is practicable, and if the plastics are in the form of containers that have held food or pharmaceutical products they may be contaminated. Not even 'green' petrol is as wholesome as environmentalists may suppose. Compounds such as benzene, used instead of tetraethyl lead to increase the octane number in some fuels, can cause cancer and reduced fuel efficiency leads to increased emissions of pollutants other than lead.

It is true that the incidence of cancers is increasing in many countries. This is due to the fact that more people are living to old age and cancer is a disease of old age. When the figures are adjusted for the age of cancer victims, the incidence of many cancers is actually falling, the principal exceptions being cancers of the lung, melanomas and non-Hodgkin's lymphoma. According to Dr Ames, 'there is no persuasive evidence from either epidemiology or toxicology that pollution is a significant cause of cancer for the general population.'[9] The World Health Organization has reported that at present there are about 14 million cancer patients in the world and incidence of the disease is expected to rise, 'the main reasons being tobacco use and,

155

paradoxically, better health care – people live longer, therefore they stand a greater chance of developing cancer'.[10]

As for more general pollution, the industrial cities of western Europe and North America are much cleaner today than they have been for a very long time: in the case of Europe for centuries. Pollution is not a new phenomenon and neither is concern over it. In 1578, Queen Elizabeth stayed out of London because of its 'noisome smells', and by the eighteenth century the air of all large British cities was severely polluted, mainly by smoke. There was legislation to reduce pollution of the Thames during the reign of Richard II in the fourteenth century.[11] The UN Environment Programme and World Health Organization monitored air pollution levels for fifteen years in 20 cities that have, or will have by the end of the century, populations of 10 million or more. Their conclusions, published in December 1992, were that pollution is serious in many of them, but of the six pollutants studied, the cleanest cities in respect of four or five of them were London, New York and Tokyo.[12]

I remember as a boy the 'pea-soupers' in the industrial city where I lived, when school closed early to give us time to get home, and I remember the last of them in London. I have ridden on a bus guided by a passenger walking in front, illuminated by the headlights so the driver could see him, and guided in turn by another passenger, a few feet in front of him, holding a fleetingly white handkerchief. Breathing was difficult and even when there was no fog, washing hung outside to dry often collected black specks of soot. The word 'smog' was coined in 1905 to describe those filthy, unwholesome mixtures of fog and smoke. Once a familiar feature of every winter, they now belong in the past. They have gone, we must hope, for ever.

We are all aware things have improved. We need only take note of the increasing concern about the cost of health care and pensions for the large number of people now surviving into their seventies and beyond to realize that more of us are surviving into old age. If fewer people are dying prematurely, the population as a whole must be healthier. Rivers, too, are cleaner than they were. Factories are now required to install equipment to remove pollutants before discharge. This is no excuse for complacency. There are still improvements that could be made and the benefits achieved in Europe and North America have not yet extended to the expanding cities of some newly industrializing countries. The solution to that problem lies in technological assistance, encouragement and further economic development to generate the capital for investment in more advanced and cleaner industrial plant.

Environmental pollution is objectionable, but it is ridiculous to suggest it

156

threatens the survival of humans or any other species. Some of those who exaggerate risks are well intentioned. They hope to 'raise public awareness' in order to achieve further improvements and they may even believe the tales they tell. Others may have other motives, based on a fundamental opposition to industrial manufacturing of almost any kind, a view rooted in nostalgia for the pastoral Golden Age they suppose us to have lost and disdain for the frivolities of modern life. In particular, they seem dedicated to opposing economic development in countries poorer than their own, partly because they maintain that people will be happier and healthier if they remain poor and partly because they fear that such development would harm the world as a whole.

We pay a high price for such indulgence. The removal of pollutants from gaseous emissions and liquid discharges is not free. It requires advanced scientific techniques for monitoring and equipment to capture and retain the pollutants. The observance of limits on the concentrations of pesticide residues permitted in our food and water also requires monitoring equipment and possibly changes in agricultural practices that increase the cost of food. The cost is related directly to the proportion of each pollutant we seek to remove, and for the last few per cent it is very expensive indeed. We should consider, therefore, whether we are prepared to meet the cost of removing substances that are unlikely to harm anyone. After all, the world is a dangerous place. Most of the chemical compounds that kill insects and fungi are produced by plants themselves, as a form of defence, and we eat them every day without noticeable ill-effect. The most toxic substances occurring in the environment do not come from our machines, but from living organisms.

There is another, even higher price we pay for our exaggerated fears. Beyond a certain point, the imposition of ever stricter controls inhibits innovation. It becomes too expensive and too difficult, and it takes too long, to submit a new product to all the tests it must pass before its use can be licensed. This may please some people, who suppose innovation to have gone too far already, but they should pause, for it may inhibit the substitution of a safer product for one more harmful. DDT, which is now largely banned, replaced very much more poisonous and no less persistent insecticides based on lead and arsenic.

Should the prognostications of doom be correct, then clearly we stand condemned and have been justly sentenced to extinction. The extremists leave us no hope, no possibility of a future, but they do so by ignoring or denying recent history and those facts that refuse their thesis. Should we

157

prefer to believe that a future is not only possible, but could bring marked improvement in the quality of human life, we should start by recognizing that substantial progress has already been made. To accept the feasibility of progress based on a more complete and more widely comprehended scientific description of the world that we inhabit can restore to us the hope that otherwise is denied.

15

The Malthusian Doctrine, or Blaming the Victim

As everyone knows, there are far too many people in the world. Or at least, there soon will be too many people, because the rate of population increase is running out of control. We are, as they say, breeding like rabbits, and rabbits are widely recognized as vermin. People are possibly the most serious of all pollutants. We hear this so often, from such distinguished persons, that it must be true. So we accept it.

The arithmetic tells its own story and never better than in the words of Paul Ehrlich, whose Population Bomb drew the matter to our attention more than twenty years ago. 'Let's examine what might happen on the absurd assumption that the population continued to double every 37 years into the indefinite future. If growth continued at that rate for about 900 years, there would be some 60,000,000,000,000,000 people on the face of the earth. Sixty million billion people. This is about 100 persons for each square yard of the Earth's surface, land and sea.'[1] Images were generated in which the entire visible universe consisted of a solid mass of human bodies.

It is not really the fault of us western Europeans, Australasians and North Americans. It is not in our countries that the increase is taking place. Growth is concentrated in the less industrialized countries, among the poor. The people of whom we have, or will have, too many are poor, many of them are very poor indeed, and there seems to be little that can be done about it, beyond urging them to behave more responsibly and criticizing those who, like the Roman Catholic church, oppose the widespread provision of contraceptive devices. The choice we face in our arithmetic of poor people is between reducing one or other of the two apparent variables: the people or their poverty. In fact it is not that simple, for reducing poverty would surely encourage people to breed faster, so eliminating what we had helped them to gain at no small cost to ourselves. No matter how you look at it, the future seems pretty grim.

Obviously, such a rate of growth cannot be sustained. The question centres not on whether it will end, but how, and most commentators, following the line of reasoning popularized by Paul Ehrlich, suppose the end will be calamitous. Our numbers will be reduced by the apocalyptic old horsemen: sword, fire, famine and pestilence will sweep the world on an unprecedented scale.

Is there an answer?

Some of the problems we perceive are real. In a demographically stable population, the number of births each year is more or less equal to the number of deaths. This means not only that the size of the population remains constant, but so does its structure: the proportion of people within each age group. If a population is increasing it can only mean that the number of births exceeds the number of deaths. This skews the population structure by increasing the proportion of young people and that is what has been happening in many countries of Latin America, Africa and Asia. It creates a problem because children must be supported, which is an expense the burden of which must be borne by the adult population, the people who are economically active. If that burden proves too great, resources must be used that might otherwise be invested in economically productive activities and in a rapidly increasing population this could inhibit economic development. This scenario tends to perpetuate poverty and it is among poor people that the need for children, usually sons, to provide for their parents in old age encourages high birth rates in societies where infant and childhood mortality has traditionally been high. It explains the link between rapid population growth and poverty. Also, the process acquires a kind of momentum. As the increased numbers of infants grow up, they enter the reproductive periods of their own lives, and as more people reproduce, more babies are added to the total. Growth can accelerate as it becomes inherent in the population structure.

The solution is very well known. Growth begins with a reduction in mortality, especially perinatal, infant and childhood mortality. It is achieved through improvements in nutrition, sanitation and health care and is reflected in an increase in the average expectation of life at birth. If that improvement is maintained, parents come to realize that a larger proportion of their babies survive; fewer need to be born to provide the requisite number of children to ensure security for the old age of their parents. If, at this point, education is made compulsory for all children, limiting the hours they can work to help support the family, large families become economically burdensome to the parents. The next step is to provide educational and employment opportunities for women. Offered a viable alternative to repeated childbirth and motherhood and the chance to contribute directly to the family budget through paid work outside the home, many women are glad to accept, along with the enhanced social life and status that goes with it. Assuming the existence of adequate contraceptive advice and materials, the stage is then set for a slowing of the rate of population increase.

Programmes to achieve these aims have been implemented in many countries and they are known to work, but there is a snag. Implementation requires an initial investment that may exceed the capacity of a poor country. Help is necessary, and it can come only from wealthier countries, as direct funding for particular programmes, investment in manufacturing and other industries, and liberalization of trade to allow products from poor countries easier access to markets in wealthier ones, with the opportunity to earn foreign exchange. These are among the goals that may be achieved through the reform of world trade agreed in 1993, at the completion of the Uruguay Round of negotiations of the General Agreement on Tariffs and Trade (GATT).

Despite the notorious reluctance of the rich to help the poor, many countries have made remarkable progress. In Thailand, for example, the average number of children born to each woman was 6.4 in 1960 and is now 2.3, whereas in Asia as a whole the figure for 1992 was 3.3, or 3.9 if China (where it is 2.2) is excluded.[2] In Latin America the corresponding figure was 3.5. In the world as a whole, the annual rate of population growth has decreased from about 2 per cent in 1973 to 1.68 per cent in 1992 and appears to be still falling.[3]

The demographic changes the world has experienced over the last half-century can therefore justifiably be looked at in a positive rather than negative light. The increase in the rate of growth was due to increased life expectancy, brought about mainly through better nutrition and sanitation. The present decline in the rate of growth indicates further improvement as more and more countries enter the second stage of what must be seen as a transition from high-birth high-death rates to low-birth low-death rates. This is the process, known to demographers as the 'demographic transition', that the present industrialized countries passed through much more slowly.[4] To see in this a cause for complaint or to cite it as compelling evidence of the criminal irresponsibility of the human race in general and of the poor in particular is perverse. It is also degrading in its demonization of the poor, feckless, and racist in that the problem, it is held, resides among people i countries other than one's own, many of whom are not white.

All our fears are based on a misunderstanding of growth rates an ignorance of the political agenda on which the doom-laden Malthusia arguments are based. Population growth, we are often told, is proceedin exponentially. An exponential rate of growth (or decline) is one in which th increase in each period is calculated on the sum of the original value and th total amount of increment accrued during preceding periods. It is the metho used in financial institutions to calculate compound interest and its effects ca be dramatic. Represented graphically, at first the curve is almost horizonta Then it begins to rise, at an increasing angle, until it is almost vertical. Draw graph of human numbers over the last few thousand years, so far as they ca be estimated, and the resulting curve has every appearance of bein exponential.

All the widely held assumptions about population are based on th exponential nature of its increase and unwarranted conclusions about i consequences. Among non-human populations, such a rate of growth is we known, and so is what happens next. There are two alternatives, known b ecologists as the J-shaped, or density-independent, and S-shaped, or densit dependent, growth curves. J- shaped curves occur when a species is presente with abundant resources. Numbers increase rapidly until a barrier is reache through loss of the resource, then the population collapses suddenly. Th increase is independent of the density of the population. This type of growt occurs, for example, among subarctic species living in an environment wher food supplies are subject to wide variation: numbers increase rapidly in th good times, only to collapse when bad times return. The animals are nc being feckless; their breeding strategy is evolutionarily stable and allow them to exploit efficiently the resources available to them. The S-shaped curv begins with a similarly rapid growth, but as population density increases s does resistance to further growth. Numbers usually decline until they reach stable level, at which they remain constant. This type of growth is usuall associated with more stable environments.

The pessimistic assumption is that the human population is following th J-shaped curve and, therefore, that it will collapse suddenly. There is not th slightest evidence to support this contention and good reason to doubt i Collapse is triggered by the disappearance of the same resource that fuelle the growth. This may be due to its exhaustion or to an environmental change such as the onset of winter or a dry season, that removes it. There is no reaso

162

to suppose that any resource on which humans depend is liable to disappear so abruptly. On the contrary, it may be that, beyond a certain level, the density of the population discourages further increase. There may be a shortage of housing, for example, or of employment so that partners must separate to allow the bread-winners to find work. Constraints such as these, which are indeed plausible, will produce an S-shaped curve. Nothing dramatic will necessarily happen. People will not ravage the countryside tearing crops from the ground in their desperate search of food, leading to vast famines. There will be no great pandemics like the Great Pestilence, later called the Black Death, which devastated the medieval population. Birth rates will simply decline and gradually numbers will stabilize.

The Malthusian legacy

We owe most of our fears about human numbers to Thomas Robert Malthus (1766–1834). His theory was stated very simply: 'Population, when unchecked, increases in a geometrical ratio. Subsistence increases only in an arithmetical ratio . . . By that law of nature which makes food necessary to the life of man, the effects of these two unequal powers must be kept equal. This implies a strong and constantly operating check on population from the difficulty of subsistence.' The checks he identified were 'misery' and 'vice'[5] and, significantly, he found them 'confined chiefly, though perhaps not solely, to the lowest orders of society'.[6] This demonstrated, in his view, that attempts to improve living standards for the poor were bound to prove self-defeating, since they would lead inevitably to further increase in numbers and renewed pressure on resources. This is precisely the argument used today.

Malthus explained carefully the difference between geometrical, or exponential, growth and arithmetical, or 'simple interest', growth. In the first, numbers increase by doubling, as when two parents produce four (surviving) children, the four children produce eight offspring, and so on. In the second, increments are added singly. If, say, there are 100 hectares of land growing food, adding 10 more brings the total to 110, but adding a further 10 brings it to only 120. Clearly, the second rate of increase, 9.1 per cent, is slower than the first (10 per cent), and equally clearly its continuation is limited by the availability of cultivable land or other physical resources.

Anyone can understand the arithmetic, and that is the trouble; it is a great deal too simple. Malthus based his entire thesis on a ludicrously large rate of population growth, which he derived largely from the experience of colonists in North America. He reported that in 1760 in New England generally, the European population was doubling every 25 years, in New Jersey every 22 years, giving growth rates of about 2.8 and 3.2 per cent respectively, and in Rhode Island growth was even faster. Estimates can be made of the rate of population increase in England and Wales during the period Malthus described, though the first full census was not compiled until 1801. In 1761 the population increased by 0.4 per cent and by the end of the century it was growing by about one per cent annually.[7] This was very much slower than Malthus claimed.

At the same time, the availability of resources increased much more rapidly than his simple arithmetic allowed, because he restricted the term to physical resources and made no allowance for the increasing efficiency with which they might be used. The late eighteenth century was a time of rapid agricultural improvement throughout western Europe and the industrialization of the British economy generated the wealth to import from overseas food that could not be grown at home. Contrary to his predictions, population growth decreased as general standards of living improved. We have not the slightest reason to suppose that the British experience, repeated in North America and throughout most of Europe, cannot be replicated elsewhere in the world.

Malthus's thesis contains an obvious contradiction. If populations invariably increase to the limit of the resources available to them, growth among the poor should remain low, while the rich should multiply until they impoverish themselves. In *An Inquiry into the Nature and Causes of the Wealth of Nations*, published in 1776, more than twenty years before Malthus published his work, Adam Smith (1723–90) wrote: 'A half-starved Highland woman frequently bears more than twenty children, while a pampered fine lady is often incapable of bearing any, and is generally exhausted by two or three . . . Luxury in the fair sex, while it inflames perhaps the passion for enjoyment, seems always to weaken, and frequently to destroy altogether, the powers of generation.' Clearly, Smith and Malthus cannot both be correct.

From an evolutionary point of view, we might expect that the higher the social status of the male, the greater the number of females to whom he has access and, therefore, the more children he will father, the corollary being that, given a free choice, females will prefer males of high rank. Whether this rule operates in modern societies is uncertain, but there is clear evidence that it did in imperial Rome and among Mormons in the last century. It has also been recorded among the Yanomamö people of South America. The

Yanomamö are extremely warlike and the status of a man is directly related to the number of enemies he has killed; those who have killed the largest number father the largest number of children, suggesting to some scientists that this social custom may selectively breed for aggressive males.[8] Taken at face value, then, the Malthusian argument leads to economic egalitarianism – the very opposite of the idea that 'the poor are always with us', which Malthus maintained.

Thus, Malthusianism is self-contradictory and wrong. However, Malthus was engaged in more than demographic theory. He had a political point to make, and he emphasized it strongly. The full title of his Essay concludes 'with remarks on the speculations of Mr Godwin, M Condorcet, and other writers'. William Godwin (1756–1836), father of Mary Wollstonecraft Shelley and father-in-law of Percy Bysshe Shelley, firmly believed it possible to construct a society based on social and economic equality. Like many people at the time, he was inspired by the idealism of the French Revolution.[9] The Marquis de Condorcet (1743–94) took an active part in the Revolution, and he contributed by devising the educational system that was adopted throughout France. Thomas Malthus's father, Daniel, moved in these circles, was friendly with David Hume and Jean-Jacques Rousseau, and was much impressed by utopian ideas, especially those contained in Godwin's An Enquiry concerning Political Justice, published in 1793, and Condorcet's Esquisse d'un Tableau Historique des Progrès de l'Esprit Humain, published in English in 1795. Thomas sought to refute them, not by advocating an alternative political philosophy, but by seeking to demonstrate that they were founded upon a mathematical impossibility.

The Essay aroused immediate interest and quickly became influential. Around the turn of the century, English politicians were discussing an amendment to the Poor Laws that would establish a minimum wage. William Pitt (the younger) first supported but then criticized this, apparently on the Malthusian ground that it would encourage large families. Eventually, even Samuel Whitbread, who had proposed it, was converted. He withdrew his measure, believing that Malthus had clearly explained why the poor laws produced more wretchedness than they relieved.[10]

Malthus had proposed an idea, based on mathematics, so simple as to be both immediately comprehensible to anyone and apparently convincing. But essentially it enshrined nothing more substantial than his own dislike of egalitarianism and preference for the existing social order. The combination was powerful, and in later editions of the Essay Malthus emphasized ever more strongly his opposition to helping the poor, denying that they had any right

to relief. Once he had added 'moral restraint' to his 'checks', in unacknow-
ledged response to the criticism by Godwin, his case seemed even stronger. If
the Poor Law awarded allowances for children, moral constraint could have
little effect; the remedy was to abolish all allowances. The Malthusian thesis
was used then, and has been ever since, to blame the poor for their own
misfortune and to justify abandoning them to their fate. In *A reply to the Essay on
Population*, published in 1807, William Hazlitt wrote: 'The poor labour under a
natural stigma; they are *naturally* despised. Their interests are at best but coldly
and remotely felt by the other classes of society. Mr Malthus's book has done
all that was wanting to increase this indifference and apathy.'[11]

Nothing has changed. Certainly, a rapid increase in a human population
creates difficulties. The means to their resolution is well and widely known
and its implementation also serves to stabilize numbers. The only ingredients
lacking are the courage and will to proceed. Both are sapped by those who
employ Malthusian arithmetic, and since that arithmetic is so deeply flawed
and the thesis so clearly contradicted by historical fact, we can only assume
that that is their intention. People who say that the introduction of improved
health care, better education and faster economic development will lead to
inevitable catastrophe do so in order to preserve the status quo. They look to
the past, invoking a dream of times when the world was a better place, the
rich could enjoy their wealth in comfort, and the poor, especially the non-
white poor, remained discreetly hidden. Not only is this disgraceful but
since the poor are unlikely to acquiesce in it, it is also impossible to attain. We
would do better to embrace the possibility of progress towards a better
future, based on scientific findings and arguments, rather than Malthusian
pseudo-science and prejudice.

16

Food, Famine and the Depletion of Resources

There is much debate nowadays about the purpose of zoos and how they should be financed. In the midst of arguments concerning the ethical implications of keeping wild animals in captivity for public entertainment, and the advantages and disadvantages of breeding animals for later release into the wild, one of the original aims is almost forgotten. A Prospectus of the Zoological Society of London, published in 1825, stated that:

> When it is considered how few amongst the immense variety of animated beings have been hitherto applied to the uses of Man, and that most of those which have been domesticated or subdued belong to the early periods of society, and to the efforts of savage or uncultured nations, it is impossible not to hope for many new, brilliant and useful results in the same field, by the application of the wealth, ingenuity, and varied resources of a civilized people.[1]

The Society was founded partly for the purpose of identifying new species for domestication and then developing them as sources of food.

The introduction of species to Britain and their adaptation to British conditions was called 'acclimatization' and in the middle of the nineteenth century it had the support of most leading scientists. Some took it further than others. Sir William Buckland (1784–1856), Dean of Westminster, the first professor of geology at the University of Oxford, and renowned eccentric, would eat almost anything. John Ruskin once complained that a compelling engagement caused him to miss a dish of toasted mice in the Buckland household. Buckland's no less eccentric son, Frank, inherited his father's tastes, once telling a friend that he found earwigs very bitter.

Fear of famine

The Bucklands may have been extreme, but concern about the nutritional status of working people was widespread, the dread of famine real and the hope of augmenting the food supply through novel means entirely understandable. Food prices fluctuated wildly and so did wages, so the same person might be affluent at one time, while wages were high and prices low, and reduced almost to starvation a few weeks later, when prices rose and the factory was working half-time. It was also an era of great change, during which people became increasingly dependent on food prepared professionally rather than at home, a development due partly to overcrowded living conditions with woefully inadequate facilities for cooking.[2] Desperate need led to political instability. In 1830–1 there were outbreaks of incendiarism in East Anglia, presenting a threat of revolt by the poor, and the Irish Famine of the 1840s may be said to have altered the course of history. Its immediate cause was infestation of the potato crop by *Phytophthora infestans*, the fungus responsible for late blight of potatoes and favoured by the weather conditions, but it was the artificially maintained price of wheat that had made the Irish poor over-reliant on potatoes as a staple.

Malthusian arithmetic worked its magic then as it does now, but in those days it was interpreted in the absence of reliable statistics. It was fear of widespread starvation in Britain that exercised thinking people in the middle of the last century, but extrapolate their concerns to present-day Africa, Asia and Latin America and old arguments can be recapitulated. 'If the world's population continues to expand at the present rate . . . within 120 years the current production of foodstuffs will have to be increased eightfold if the standards of now are to be maintained – and yet these are inadequate for more than half of the people now living on earth,' wrote Georg Borgstrom in 1972, having already stated that 'the human race long ago exceeded the limit of what the world can feed'.[3]

Twenty-odd years ago I was as troubled as anyone about the spectre of famine spreading across entire continents and I wrote many articles and a book on the subject. The idea was widely accepted and the facts seemed to support it. Hunger was widespread in low-latitude countries, as it still is. The number of mouths was increasing rapidly, so the situation was more likely to deteriorate than to improve. In 1974, the Food and Agriculture Organization of the United Nations reported that at least 460 million people, about 11.5 per cent of the global population at that time, were seriously ill-fed and that between 1952 and 1972, food production increased more slowly than population in one-quarter of all the countries in the world.[4]

To many of us, transferring the intensive agricultural techniques used in industrialized, temperate-climate countries to poor, tropical and subtropical countries seemed impracticable and inappropriate. It was impracticable because the techniques relied heavily on machinery, fertilizers and pesticides, which were too expensive for subsistence farmers, and inappropriate because of the damage it was believed they would cause to soils and hence to the sustained production of adequate crop yields. The situation was exacerbated by the growing of crops for export, apparently denying local people the land on which to grow food. It is hardly surprising, then, that I and the many who thought as I did, phrased our fundamental question in purely Malthusian terms. We wondered how many people the planet could feed and suspected we might be perilously close to an absolute upper limit. Disagreements turned on this question: pessimists like Borgstrom maintained that the limit had already been exceeded and a catastrophic population collapse was imminent, while optimists held that sufficient time remained for us to rescue ourselves by remedial steps which centred on stabilizing our numbers.

Since then, many years have passed, during which the human population has increased from around 4 billion in 1974 to 5.4 billion in 1992. In December 1991, the International Conference on Nutrition noted that some 700 million people, 12.9 per cent of the total, did not receive enough food to meet their basic needs. At first glance our fears would seem to have been justified, even though the worst of them have not been realized. There have been famines, but none that gripped entire continents, and the figures may mislead. They show a slight decline in the amount of food per person, but imply a huge increase in the amount of food produced. The world is now feeding one-third more people than in 1974. That so many people still go hungry is tragic and disgraces us, but that the number of hungry people is not very much larger is a triumph. Like all Malthusians, our arithmetic had led us to suppose a world that was much simpler than it really is. In particular, we saw an essentially agricultural problem, when the reasons for famine and chronic hunger are seldom agricultural.

The causes of famine

Famine occurs when harvests fail and those dependent on them are unable to obtain food from elsewhere. Harvest failures are most commonly due to adverse weather, but almost as commonly they are due to war, which drives rural populations from their land and leaves crops rotting in the fields. When crops fail or are lost, people either import food or move away to where it is more plentiful. The poor cannot afford to pay for imports and, in the modern world, large-scale migrations are discouraged by guarded international frontiers. Those who manage to cross usually find themselves repatriated or confined helplessly in camps. The traditional solution to hunger, used throughout history, especially by the peoples of marginal lands such as those bordering deserts, has been effectively removed. Imported food has to be paid for, causing food prices to rise, which encourages hoarding and speculation, and invariably it is the poorest who are the hungriest.

When we see pictures of hungry people it is easy to confuse two distinct phenomena and to assume that the underlying cause of the famine is overcrowding, that people are hungry because there are more of them than the land can support. This is not so, though poor practices, such as over-grazing by the herds on which pastoral peoples depend, can aggravate difficulties that already exist. Marginal lands, such as those of the Sahel region of Africa, are subject to periodic droughts, which can persist for several years. When they end, which invariably they do, plants grow once more. Wars, even those of our own century, also reach conclusions, allowing farmers to return to their lands and resume cultivation.

Acute famine can be relieved by food supplied free of charge from other countries, but chronic shortage cannot. The strategy was tried many years ago and it failed. Food surpluses from elsewhere in the world, such as Europe or North America, can be provided either free or merely for the cost of transport. When it arrives, however, the free or almost free food undercuts prices and, therefore, the income of local farmers, leading to a reduction in local production. Despite the present embarrassment of our huge surpluses, they are irrelevant to the needs of the chronically hungry, although they can help the famine-stricken.

Famine creates a disaster, but a temporary one. It is not the result of the inadequacy of farmers and neither is chronic hunger. At one time, the great majority of people lived in the countryside and most were directly concerned with food production. The food that they grew was consumed locally. This is the situation some consider ideal, though it is doubtful whether they have experienced personally the extremely low living standards and oppressive social environment associated with it. As economies begin to develop, people are drawn to the towns and must buy the food they can no longer grow for themselves. In this respect, the situation in the less-industrialized world is similar to the one that obtained in Europe during its transition to an industrialized economy. That is why modern concerns about the world food situation so closely parallel concerns about the national food situation in Britain in the last century.

Then, the difficulties were partly logistical. When most people ate locally grown food, so far as Britain was concerned there was little need for transport except into London. As other cities grew, they also required food transport and at first this was wholly inadequate for the task. Roads were in a very poor state and waterways were little used to convey food. Roads were improved, the railway network was built, refrigeration became available, and it became much easier and cheaper to supply perishable foodstuffs to urban populations. More recently, economic support mechanisms were devised to stabilize prices. The strategies were more than successful: they have led to the present European food surpluses. Identical difficulties have occurred in the countries of Africa, Asia and Latin America; the solution is the same, with the important difference that the necessary technologies are now well known.

Once people have to buy their food, obviously they need to have the money to pay for it. Equally, farmers must be able to sell their food. The supply and demand sides of the equation will find a balance. If supply exceeds demand, prices fall, leading to a reduction in future supply, which restores prices. Both producers and less affluent consumers suffer through price instabilities. As soon as the logistical difficulties had been overcome, hunger became a symptom of poverty for consumers and of price instability as it affected farmers. The alleviation of poverty was a social and economic problem to be solved through political reform, aimed mainly at stimulating the demand for food. It is one of the more curious paradoxes that where farm output is low and hunger is rife, the underlying cause is almost invariably a lack of effective demand for food.

The green revolution

Hunger, then, is not primarily an agricultural problem and cannot be solved simply by increasing farm output. The demand must come first, although agricultural improvements may be necessary to ensure that farmers are equipped to respond to it. From the eighteenth century, European and North American farmers made radical changes to their methods, and these are continuing. Even more dramatic advances have been made during the second half of this century in low-latitude countries, through the complex of programmes nicknamed the 'green revolution'.

These programmes began with a scientific advance. The most obvious way to increase crop yields is to apply fertilizer. This makes plants grow larger, but where the crop is the seed, rather than leaves or the whole plant, this may not be desirable. A tall plant may become mechanically unstable and liable to fall, making it more difficult, or even impossible, to harvest. The need, then, was to develop plants that would grow well in low latitudes and respond to increased fertilizer applications by producing more seeds, but on a larger number of shorter stems. Wheat varieties meeting this requirement were developed at the Centro Internacional de Mejoramiento de Maiz y Trigo, in Mexico, by a team led by Dr Norman E. Borlaug, who was awarded the 1970 Nobel Peace Prize for the achievement. The first of these new varieties became available commercially in 1962. It yielded 15 to 20 tonnes per hectare, compared with about 8 tonnes for traditional varieties. It was introduced rapidly in many countries. In India wheat production tripled between 1966 and 1979. At the International Rice Research Institute, in the Philippines, a programme to develop new varieties of rice began in the early 1960s. The first, called IR-8, was introduced commercially in 1966. Not only did it produce more grain, it also grew more rapidly, allowing two or three crops a year to be grown on land that previously supported one or two, and increasing output from less than 2 tonnes per hectare to as much as 16.

The new high-yielding varieties required fertilizer in order to achieve their full potential, although without it they will produce crops that are smaller but still larger than those from the varieties they replaced. Pesticides were needed to reduce losses in the field, as were improved storage facilities. In low latitudes, up to half a crop can be lost to pests and fungal contamination after it has been harvested. Larger volumes of materials called for better transport and farmers needed better credit and banking arrangements as their subsistence farms were transformed into small businesses.

Scientists and technologists provided the means to avert the threatened world food crisis and they continue to do so. Traditional plant breeding is giving way to faster, more precise genetic manipulation (see chapter 22), for example, to produce plants resistant to insect or fungal attack and with better storage properties, while reducing reliance on pesticides. Agronomists have devised irrigation systems that use water much more efficiently. Highly trained advisers work alongside farmers to study their methods and suggest the small improvements that can dramatically increase the amount of food they produce. Should agriculture prove incapable of producing enough, scientists have also devised many less conventional ways to produce highly nutritious food. You have probably used them, as 'TV' (i.e. textured vegetable protein) dinners or substitutes for milk that were never within miles of a cow, and there are more. Meanwhile, it is true that land is devoted to the growing of crops for export. This does not necessarily deprive people of food if it provides employment and generates wealth that can be invested to provide still more jobs. When people have money they will buy the food they need and there is no evidence that would lead us to suppose that farmers are incapable of supplying it.

The original question must remain unanswered. No one can know how many people the world might feed. All we can say is that if a limit exists it is not yet in sight and, as I pointed out in the last chapter, the human population seems to be following an S-shaped growth curve which is leading it towards stability. From this point of view, as from every other, the future is very far from bleak. Of course, things could go wrong. Should the climate change drastically there might be serious agricultural consequences, at least until farmers had adjusted to the new conditions. But even then, we should expect gains in one area to compensate for losses in another.

The principal problem remains economic. The cycle is very simple: industry provides jobs, which pay wages, so that there is money to buy food, providing farmers with the return that they need to grow crops, and governments smooth the process by stabilizing markets and regulating the supply of credit. That is how it was resolved in the industrialized countries. So far as we know, it is the only way it can be resolved and those who oppose economic and industrial development seek only to perpetuate the hunger and poverty they pretend to abhor. Provided we allow that it is possible to solve the problem we perceive, it can be solved. All we need is confidence in the future, a belief in the feasibility of progress, and determination to support those scientists who are working to develop the new crops and new methods that are fated eventually to become as hallowed and traditional as the horse-drawn plough.

Using up resources

The availability of food depends on its annual production. It is a renewable resource and hence flexible. Non-renewable resources are not like this. The amount of them available to us is fixed and once we have used them up they will be gone for ever. How real is the risk that the world will exhaust its supplies of materials on which our economies depend?

173

I live in Cornwall, in the far south-west of mainland Britain. Today tourism is probably our most important industry, but traditionally the region was sustained by farming, fishing and mining, with workers moving fairly freely among the three according to the dictates of economic conditions. Farming and fishing continue, but mining is now restricted to the extraction and processing of china clay (kaolin). Hard-rock mining has almost completely ceased. The miners used to extract a wide range of metallic ores. Many mines were deep and some extended beneath the seabed. The work was harder and more dangerous than coal mining; they used to say a miner was lucky to survive past his thirty-fifth birthday. Mining began many centuries ago and metals, especially tin, were exported from Cornwall in Roman times. Tin was the most important local resource and an essential one when tools and weapons were made from bronze, an alloy of copper and tin. Cornwall also produced copper.

It is all finished now. Cornish mines could not compete with cheaper tin mined in other parts of the world and a world-wide glut of tin contributed to the failure of the international mechanism that had previously regulated the market. One or two mines remain open as tourist attractions and one still produces tin commercially, but the industry was destroyed by the collapse in world prices for its products. Tin mines closed, essentially because the world had more tin than it needed or could use. This should not have been the reason. Every tin mine in the world should have closed a few years ago because it had been predicted that, by around 1987, the world would have run out of tin. About now, according to the same set of predictions, all the copper mines should be closing, their reserves exhausted. We should have run out of mercury in about 1985, and ten years from now we should run out of aluminium. Eventually, of course, we are predicted to run out of more or less everything on which our way of life depends. They are non-renewable resources and once consumed they are gone for ever and can continue in use only by recycling.

These predictions were made in one of the first attempts to study the world holistically by constructing a mathematical model using what was then one of the most powerful computers in the world, at the Massachusetts Institute of Technology. The study was commissioned by the Club of Rome, an 'invisible college' of thirty distinguished individuals convened by Dr Aurelio Peccei, an industrial manager and economist, and charged with investigating 'the present and future predicament of mankind.'[5] The results were published in 1972 as The Limits to Growth and, not surprisingly, they caused widespread controversy.

The model was designed to investigate five areas: industrialization, population growth, malnutrition, depletion of non-renewable resources and environmental deterioration. It made certain assumptions, for example, that increasing pollution leads to increased mortality, and concluded that: 'If the present growth trends in world population, industrialization, pollution, food production, and resource depletion continue unchanged, the limits to growth on this planet will be reached sometime within the next one hundred years. The most probable result will be a rather sudden and uncontrollable decline in both population and industrial capacity.'[6] Critics pointed out that many of the assumptions were flawed. Rising pollution, for example, does not necessarily lead to increased mortality, and mineral resources might not be depleted in quite the way the model suggested.

Collapse through the exhaustion of metals was not altogether a new idea. The American geologist Preston Cloud had been warning for some time that mineral resources are finite and likely to be exhausted in the near future. After listing 20 of the most important minerals, he predicted that only 11 would last beyond the end of the century,[7] and his warning was repeated by many authors concerned at what they saw as the threatening catastrophe. Paul and Anne Ehrlich wrote that: 'Even if world population growth stopped in 1972, world iron production would have to be increased about sixfold, copper production almost sixfold and lead production about eightfold to bring global per capita consumption to the present United States level.'[8] Throughout the many books and articles written on the subject, the concern over the fate of the United States was paramount, for its days of self-sufficiency in oil and most metals were drawing to a close. Americans were becoming importers, from a world that seemed suddenly unreliable. There were warnings of 'resource wars' and, of course, the United States did go to war with Iraq to protect its oil supplies.

All the calculations were based, of course, on the old Malthusian trick of exponential growth, or in this case depletion, and the clear implication was always that the main key to resolving the problem and conserving essential supplies would be found in restricting population growth among the poor. The scale of our difficulty was demonstrated by graphs comparing 'static' and 'exponential' reserves. The 'static' reserve, assuming continued resource exploitation at the present rate, shows that supplies of metals last long into the future; the 'exponential' reserve, calculated at an exponentially increasing rate of exploitation, shows a range of them exhausted before 1990. The demand for oil was predicted to exceed supply during the 1990s. E.F. Schumacher, a statistician formerly employed by the National Coal Board, was wisely unimpressed by these predictions. He once said to me that 'everybody knows that if you go on inflating a balloon eventually it will burst. So what?'

Resource depletion and its link with population was also the theme of 'A Blueprint for Survival', of which I was one of the five authors. Published in January, 1972, as a special issue of *The Ecologist* magazine and subsequently in book form, it was translated into many languages.[9] The 'Blueprint' was a subtle mixture of turgid prose and apocalyptic prognostication which pulled no punches in its sensationalization, but it differed from other warnings in one important way. 'The principal defect of the industrial way of life is that it is not sustainable,' it began. 'Its termination within the lifetime of someone born today is inevitable . . . We can be certain . . . that sooner or later it will end . . . and that it will do so in one of two ways: either against our will, in a succession of famines, epidemics, social crises and wars; or because we want it to – because we wish to create a society which will not impose hardship and cruelty on our children – in a succession of thoughtful, humane and measured changes.' These recommended changes were designed to lead to small, self-contained and largely economically self-reliant communities and a rapid end to population growth throughout the world, with free and freely available access to contraception, sterilization and abortion.

A major misunderstanding

A few years later the emphasis changed. The second report to The Club of Rome dealt mainly with population growth and the 'north–south' relationship between industrialized and less-industrialized countries, but said little about mineral resources other than oil.[10] The more sophisticated report of the Brandt Commission, published in 1980, was concerned about the slow and patchy industrialization of the poorer countries and the risks of protectionism and, in its much more rational view of population trends, drew attention to the difficulties faced by migrant workers.[11] The 1987 Brundtland Report, which popularized the concept of sustainability first proposed in the 'Blueprint', was more specific about mineral resources: '. . . their use reduces the stock available for future generations. But this does not mean that such resources should not be used . . . Sustainable development requires that the rate of depletion of non-renewable resources should foreclose as few future options as possible'.[12]

Clearly, our fear that, before much longer, we would exhaust the global stock of mineral resources was mistaken: the resources have not disappeared. Our error was twofold. In the first place, we misunderstood the meaning of the words 'stock', 'reserve' and 'resource', often using them as synonyms for one another when they are used with precise and different meanings by those engaged in the relevant industries. The 'stock' of a substance is the total amount that exists on the planet (or, perhaps, in the solar system) and in most cases no one will hazard a guess at the size of that amount. Aluminium, for example, is by far the most abundant metal on our planet. The uppermost 16 kilometres of the Earth's crust is about 8 per cent aluminium by weight, a fact that should have alerted us to the unlikelihood of its exhaustion. If the substance is of industrial or other use it is a 'resource', a term that implies no quantitative value. Mining companies extract 'resources' and measure the amounts available to them as 'reserves' and it is only to the term 'reserve' that quantities can properly be attached.

The figures on which 'static' and 'exponential' depletion rates were calculated were those for reserves. They were obtained from the US Bureau of Mines, and all the forecasters used the same figures. The source is highly respectable, of course, but its data are open to misinterpretation. The Bureau compiles the figures from those it obtains from mining companies; in other words, the reserves are those notified to it by the industry. The companies obtain their reserve figures from surveys and estimates and there are different levels of reserves. Where there are firm measurements, from mines, surface rocks and boreholes, the figure is for an indicated, or proven, reserve. If the figure is an estimate based on experience and a detailed knowledge of geological structures, but with no detailed surveys, it is an inferred reserve. If the reserve is suspected, but not proven, to exist, it is a potential reserve.

All these categories are linked by the underlying definition of a reserve, as a resource that can be mined profitably and legally under existing circum-stances. In other words, the concept is entirely economic. Should the circumstances change, the figures will change accordingly. Mining surveys and explorations are costly and so companies are frugal in their use. Each company needs to know how long it can continue to work its existing mines, the location and size of the reserves to which it will move when its present mines are depleted, and the most profitable places to explore for new reserves beyond that. Mines do not become depleted by exhausting the minerals obtained from them; that, too, is an economic concept and a fluid one. It may become too expensive to continue working a mine though it is far from worked out, and a subsequent price rise may render previously uneconomic mines profitable once more.

Mines vary in size and in the length of time for which they can be worked but on average they last about thirty years. For this reason, companies tend to think about thirty years ahead. No manager is much interested in commodities that will not be sold until further in the future than that. Their reserve figures reflect this, as you might expect, so, at any particular time, the industry will appear to predict the exhaustion of reserves some thirty years ahead. It does not mean the world has enough of each mineral to last no more than thirty years, but that is how the data were misinterpreted.

In The Limits to Growth, an allowance was made for this by assuming five times the indicated reserves, but this served merely to dramatize the risks because it still failed to take proper account of the nature of the reserve concept. Should demand increase, exploration will intensify and reserves will increase, probably at least as rapidly. Between 1950 and 1970, for example, because of increased demand, bauxite reserves (the principal source of aluminium) increased by 279 per cent, copper by 179 per cent, chromite (the main chromium ore) by 675 per cent, and tin reserves by 10 per cent. [1]

Human resourcefulness

Our second mistake arose from our gross underestimate of human ingenuity, the one resource that appears to be infinitely renewable and incapable of depletion. If it seems that a resource may become scarce, there will be strong incentive to find new sources and also to discover substitutes. Sometimes, however, the ordinary course of technological development or what I choose to call 'progress', generates substitutes not as somewhat inferior replacements for scarce materials, but as improvements. Copper, for example, was once used in vast quantities for the manufacture of undersea telephone and telegraph cables and large amounts are still contained in ordinary telephone wires. Today, intercontinental telephone calls travel not under the sea, but far above it. The call travels by land line to a transmitting station, from there by radio to an orbiting satellite which relays it to a receiving station, and from the receiving station to the recipient by land line. The undersea cable is obsolete, and now land lines are being replaced by optic-fibre cables, made from glass which itself is made mainly from sand. This is not a response to an immediate or imminent shortage of copper, but because only by using optic-fibre cables will it be possible to meet the rapidly increasing demand for more advanced communications. Satellites and optic fibre cables are superior to copper cables and in this use copper is rapidly becoming obsolete. The problem now is not the risk of running out of copper, but of helping the economies of those countries and regions that are too heavily reliant on its mining and refining. Mercury was once used to make heavy-duty industrial switches and it, too, is a metal we were warned was in danger of exhaustion. Switchgear is now electronic. Small chips turn machines on and off, and mercury switches disappeared years ago.

Depletion, then, is much less serious a problem than it may appear and this is true even for commodities for which there seems to be no immediate alternative. Oil, for example, is abundant, but no one imagines we can continue to use it indefinitely at the present rate, yet it is unlikely that its disappearance would cause serious difficulties, except to countries in which it is the only source of revenue. It is no longer required for power generation, of course, and the electrification of rail lines increases greatly the number of primary fuels that can drive trains. There are alternatives, even for road transport. Electrically-powered cars and buses are in an advanced stage of development, while hydrogen can be used as a fuel that yields more energy per unit mass than petroleum, and produces only water when it burns. There are rival claims for ethanol (alcohol) and methanol as 'biomass' fuels produced from cropped plants, and a range of engine designs is available to suit these and other fuels.

Meanwhile, cars are being developed that will achieve much greater efficiency in their use of gasoline. The Lotus company is planning a car that will travel 35 kilometres on one litre of gasoline (100 miles per gallon) at high speed, and 50 kilometres (140 miles) at low and cruising speeds. General Motors' Ultralite, conceived by Amory Lovins, who was once active in Friends of the Earth, offers even more. With a revolutionary redesigned body to reduce aerodynamic drag, the car would use a 250 cc engine to drive a 10 kilowatt generator, which would charge the batteries supplying power to one electric motor for each wheel. The car should be capable of travelling 100 kilometres on one litre of gasoline (almost 300 miles per gallon).[14] Chrysler have developed the 'Patriot', a car using two gas turbine (i.e. jet) engines to generate electricity fed to the induction motor that supplies the motive power. When the motor requires less than the full turbine output, the surplus electricity is fed to a flywheel spinning in a vacuum housing, which stores energy for use as required, especially in fast acceleration. It is hoped that the Patriot will have a top speed of 320 km/h (almost 200 mph) and will run on liquefied natural gas, estimated to emit less than one gram of pollutants per kilometre. This complex machine converts fuel into useful work with twice the efficiency of a comparable internal combustion engine.

The Patriot is to be entered for the 1995 Le Mans 24-hour road race and if all goes well there, for other road races.[15] We should never underestimate not only human ingenuity, but the sheer delight humans find in solving problems and making difficulties evaporate.

Ironically, perhaps, the real problem faced by mining companies arises not from the exhaustion of their resources but from the public reaction to their plans. In the industrialized countries, any proposal to open or extend a mine will be opposed vociferously by those who object to the pollution that they fear it will cause and the damage it will inflict on the environment. Sometimes such objections can be dismissed as 'NIMBYism' (Not In My Back Yard) or even as manifestations of (the ungrammatical) 'BANANA' (Building Absolutely Nothing by Absolutely Nobody Anywhere), but some reflect genuine and rational concerns. This is especially true in Britain, a crowded island with great competition for space, and it has led to tremendous battles. In 1966, for example, it was learned that Rio Tinto Zinc planned to mine for copper in the Snowdonia National Park. The news triggered a national campaign led by Graham Searle, who became the first British director of Friends of the Earth following its formation in 1970, and the struggle lasted several years.[16] A similar battle was fought a few years earlier over plans to mine iron ore in the attractive countryside of north Oxfordshire.[17]

Such conflicts are not resolved easily and the inevitable compromise usually leaves victims. It may be that the best strategy is avoidance, and lies in encouraging scientific and technological research to develop not substitutes, but complete alternatives. Believe in the future and you may come to recognize such devices as telephones, videophones, computers and modems not as the malevolent products of scientific arrogance and technological oppression, but as alternatives to, say, some of the road journeys many business people are required to make.

Concern over resources now centres less on minerals for industry than on the provision of water for irrigation and drinking, and less on oil than on wood, which is still the most widely used fuel in the world as a whole as well as the most important building material. Attention is focused not so much on shortages of non-renewable resources, as on shortages of renewable ones. But these can be overcome, provided that there is the will to do so, through a combination of accelerated economic development and technological advance.

There is nothing new in the realization that we face challenges. Humans have always faced challenges and those civilizations that demonstrated the greatest determination in meeting them rationally and devising appropriate responses have prospered. It is wrong to bemoan the existence of challenges and to seek refuge in attempts to dismantle our economic structures in the vain hope of recreating some lost arcadia where everyone might be securely confined. If we allow it, many solutions will occur almost of their own accord, as ordinary progress substitutes new resources for old, leaving us free to concentrate our efforts on the more intractable difficulties that remain. Those efforts are most likely to succeed if we take the optimistic view that solutions are not only feasible but probable.

17

Greenhouse or Ice Age?

Geologists, who divide the history of our planet into eras, periods and epochs, say we are now living in the Holocene (or Recent) Epoch of the Pleistogene Period of the Quaternary Sub-era of the Cenozoic Era. Our epoch began about 10,000 years ago, when the last glaciation ended. At least, that is what most geologists say, but some disagree, believing it premature to describe the present epoch as 'recent'. We are living, they say, in the Flandrian Interglacial, a warm interlude between the last (Devensian) Glacial or ice age and a new one that may begin quite soon, which means we are still in the Pleistocene Epoch, the time of ice ages, and it shows no sign of ending.

Palaeoclimatologists, who study the climates of the distant past, generally share this view. They hold that we are due for a return of the glaciers and ice sheets, though the next ice age will be a minor one. There have been about twenty ice ages during the last two million years or so, occurring at intervals of around 100,000 years and separated by interglacials, each lasting about 10,000 years. It is believed that the commencement and termination of glacial episodes are triggered by cyclical changes in the Earth's rotational axis and orbit (see note 4 to chapter 4), so it is possible to calculate when the necessary conditions will recur.

Global warming

I mention this only by way of warning. Today it is the risk of global warming that concerns most people, but the calculations on which this threat are based are very complex, somewhat crude, and the issue is surrounded by considerable uncertainty. It may be that during the next century the world will become warmer as a result of the gases we are releasing into the atmosphere. On the other hand, nothing may happen, or the climate may grow cooler, or it may begin to warm and then cool.[1] No one really knows and the actual evidence, from measurements, provides little indication of what will happen. At a meeting of the Royal Meteorological Society, held in London in December 1992, Tom Wigley, one of the climatologists leading research into global warming, said that if the climate responds sluggishly, it could be thirty or forty years before there was clear evidence of it. 'I cannot stress the uncertainty enough,' he said. Sir John Mason, former director of the Meteorological Office, complained that politicians needing to make decisions were asking the scientists studying climate 'to run before we can walk; to make predictions we may regret'. The meeting was told by Dick Lindzen, professor of meteorology at Massachusetts Institute of Technology, that his calculations suggested that the doubling of the atmospheric concentration of carbon dioxide by the middle of the next century, the assumption on which all calculations are made, would increase average temperatures by 1.5°C at most, compared with the 1.5 to 4.5 degrees more usually predicted.[2]

The underlying principle is simple enough, not in the least controversial and it was discovered long ago. John Tyndall (1820–93), the British physicist who pioneered the study of heat and the scattering of light, mentioned, in a lecture delivered in the United States in the early 1870s, that nitrogen oxides absorb radiation.[3] Later, the Swedish physical chemist Svante August Arrhenius (1859–1927) pointed out in his book *Worlds in the Making*, published in 1908, the climatic implications of the absorption of radiation by carbon dioxide.

Electromagnetic radiation spans a vast range of wavelengths, with radio waves at the long-wave end, heat, visible light and ultraviolet light in the middle of the range, and X and gamma rays at the short-wave end. The Sun radiates across the full spectrum, but the Earth is bathed predominantly in heat and visible and ultraviolet light, to which the atmospheric gases are transparent. Once warmed, the Earth also radiates, but at longer wavelengths. If this outgoing, or 'black body' radiation strikes a molecule comprising more than two atoms it may be absorbed by the molecule, causing it to move more rapidly, which is another way of saying it is heated. This molecule may strike other molecules, transferring some its acquired energy to them, and so a significant portion of the energy it received from its absorbed radiation is held by gaseous molecules in the atmosphere, which are warmed.

According to their sizes and shapes, different molecules absorb at different wavelengths and, obviously, the amount of energy absorbed depends on the number of suitable molecules present. The most abundant of these molecules are those of water vapour, absorbing at wavelengths of between about 1.0–4.0, 6.0–9.0 and 25–60 μm (micrometres), and carbon dioxide, which absorbs at about 5.0 and 19–20 μm. Other gases also absorb, each at its own

183

wavelength, and their combined effect is to hold the average temperature about 38°C higher than it would be without them. This is the so-called 'greenhouse effect'. Clearly, if the concentration of 'greenhouse' gases increases, we should expect the average temperature to rise, and, since the burning of coal, oil and gas is known to be adding significantly to the amount of carbon dioxide in the air, and we are also adding a number of other relevant gases, it is reasonable to suppose that sooner or later the world will become warmer.

It is a simple message, but basically a political one. Yet again, we humans are depicted as destroyers of the planet. As the temperature rises remorselessly, crops will wither in the fields, the seas will rise to inundate low-lying coastal areas and eventually the Antarctic ice sheets will melt, adding enough water to the oceans to turn mainland Britain into a group of islands and revising the North American coastline, so that most of the cities from Houston to Boston are swamped. Never before, the campaigners protest, has the global climate changed so fast and on so vast a scale, and the guilt for this sin is ours. Could we but return to the true faith embodied in the simple life and cease the wicked oxidation of carbon with which we pander to our greed for travel and possessions, we might yet be saved. As we know all too well, every aspect of our sin is increasing at an exponential rate. An attractively coloured piece of artwork in The Gaia Atlas of Planet Management tells it all: population (pink), energy consumption (brown), number of scientific journals (green) and mobility in kilometres per hour (blue) all increase until the curves describing them rise vertically, soon to contradict most known laws of physics.[4]

Attention has been drawn to certain curious aspects of these warnings. Peter Westbroek of Leiden University in The Netherlands and several colleagues have pointed out the 'numerological significance assigned to the year 2000', a magical point at which values reach infinity, having remained virtually flat throughout the whole of pre-Renaissance history, when, presumably, everyone lived in harmony with their planet. It is, they say, 'a weird mixture of ideology and science, or fact, fiction and fear' and the graphs could be drawn differently. If, for example, there are real and reliable data on these topics for the period from 1500 to 1800, which is doubtful, the period prior to 1500, for which data have been guessed, could be omitted and the vertical scale could be greatly reduced. This would produce a much flatter and so less dramatic curve.

A history of climatic change

Westbroek and his colleagues then go on to argue an equally extreme case, but one that is well supported, to show that the activities of early pastoralists triggered the large-scale conversion of tropical rain forests into savannah grasslands, producing cooler and drier climates. It was this change, they suggest, that destabilized the global climate at the commencement of the Pleistocene, bringing about the sequence of glacial and interglacial episodes that have dominated the climate ever since. This was 'the first man-made environmental catastrophe of global extent'.[5]

The Westbroek paper is not meant to be taken too literally. It merely attempts to challenge the widely accepted idea that only recently have humans begun to influence the global climate significantly, by urging more investigation of changes that occurred long ago. This is a reasonable proposal and while the case is deliberately overstated, most scientists accept that the clearance of forest by burning played an important part in establishing most if not all of the world's major grasslands. The induction of environmental change by humans is not a recent phenomenon and our ancestors lived less ecologically sound lives than we are led to suppose.

We tend to think of the climate as constant, changing only over very long periods, but this is not so, either. It is always changing. 'When milk comes frozen home in pail' was not Shakespearean fantasy but an image that would have been all too familiar to the audiences for Love's Labour Lost, who first heard this line in the 1590s. Winters really were colder then, because much of the northern hemisphere was gripped by the 'Little Ice Age', a cold period that lasted from about 1550 to 1860, with especially low temperatures around 1600. The warming of about 1°C which most climatologists believe has occurred over the last century may be nothing more than recovery from the Little Ice Age.

It was not until 1893 that a link was discovered between this cooling and solar activity. The British astronomer Edward Walter Maunder (1851–1928) found very few reports of sunspots between 1645 and 1715 and a period of 32 years when no sunspots at all were recorded. This period became known as the 'Maunder Minimum'. In the 1970s, the American solar astronomer John A. Eddy investigated the phenomenon further. He discovered a very strong correlation between sunspot activity and global temperature stretching back to 3000 BC, with at least twelve pairs of Maunder Minima and 'Little Ice Ages' during that time.[6]

Climatic predictions are based on computer models of the general

185

circulation of the atmosphere. These are constructed on the basis of an imaginary three-dimensional grid over the entire planet, with lines a few hundred kilometres long in the horizontal plane and a few kilometres high in the vertical plane. At each intersection of grid lines, calculations are made of the behaviour of the atmosphere in response to previous changes and changes at adjacent intersections. The models are highly complex, but suffer never-theless from the disadvantage of all such holistic endeavours: they are forced to oversimplify. Cloud formation, for example, occurs on scales much smaller than those used in the grids, so it is difficult to model this accurately despite its obvious importance, and it is complicated. Clouds reflect incoming radiation, cooling the surface, but the condensation of water vapour releases latent heat, which warms the surrounding air, and much depends on the kind of clouds that form, some types having a general warming, others a general cooling influence. Water warms and cools much more slowly than land. The oceans act as an important 'heat sink', with ocean currents transporting warm water from low to high latitudes. However, their effects are not fully understood and modelling of them is necessarily imprecise. Doubts also surround the fate of the carbon dioxide entering the air, much of it disappearing somewhat mysteriously, perhaps into vegetation. The models are being improved as fast as increasing expertise and computer power allow, but many uncertainties remain and will do so for several years to come.

There is general agreement, however, that the Antarctic ice sheets are very stable. It is most unlikely that they will melt unless temperatures rise much higher than even the most pessimistic estimates suggest, though there could be a small rise in sea level as the warmed oceans expand.

When, eventually, reliable predictions do emerge, the last thing we can expect is a simple, straightforward warming. If you look forward to growing oranges in Cumbria or olives in New York you may be disappointed. The last time the world warmed, as the most recent full glaciation came to an end, the climate at first grew warmer, but then, rather suddenly, plunged back into ice-age conditions. This was signified by the appearance of a small Arctic and alpine plant, Dryas octopetala, the mountain avens, whose pollen has been found in soils of the time over much of northern Europe. The cold lasted for several centuries, gave way to resumed warming and then returned for a second time.[7]

The Dryas episodes may have been caused by changes in the North Atlantic. When sea water freezes, its salts are removed, making adjacent water more saline and therefore denser. Its density also increases because its temperature is close to 4°C, at which water density reaches its maximum.

Close to the edge of the sea ice, this water sinks beneath the less dense water and forms a current, called the North Atlantic Deep Water, flowing towards the equator, and is replaced by water flowing northwards closer to the surface, thus establishing a circulatory pattern of currents. The warm currents begin in the Gulf of Mexico as the Gulf Stream and turn south again at about the latitude of the Iberian Peninsula, except for a branch, the North Atlantic Drift, which continues northwards, passing the western coast of Britain. As the climate warmed and the ice sheets retreated, large amounts of fresh water must have poured into the ocean to float above the denser sea water. At the same time, the retreat of the edge of the sea ice may have altered the formation of the North Atlantic Deep Water. This might have changed the oceanic circulation, so the North Atlantic Drift ceased to flow northwards from the Gulf Stream, producing profound climatic change. If this is what happened before, then perhaps it might happen again under conditions of rapid warming. Interestingly, or perhaps ominously, it was reported in *The Daily Telegraph* (10 March 1994) that scientists had detected a reduction in the formation of North Atlantic Deep Water, which was apparently associated with a retreat of the edge of the sea ice.

Warmer weather is likely to be wetter weather, because evaporation will increase, and some predictions show large areas of what are now desert receiving enough moisture to sustain permanent vegetation and, therefore, agriculture. This would represent a distinct improvement, but one that would be partly offset if the deserts simply migrated further away from the equator, into the southern part of the North American grain-growing region, for example. It is all uncertain, as is the way plants might respond to warmer conditions and air containing more carbon dioxide.

So great are the uncertainties that we cannot be sure that warming will occur or, if it does, that its effects will be malign. It is always best to be prudent, so it would make sense to consider the kind of changes that may become necessary, but the campaign to impress upon us the urgency of this supposed threat may do much more harm than good. It is not the case that any action of any kind is better than no action at all. Responses must be informed, especially when the campaigners reject the one technology – nuclear power – that generates electricity reliably without emitting carbon dioxide (see the next chapter).

It is intellectually dishonest to misrepresent the state of scientific knowledge for purposes of political propaganda. Given the poor state of public understanding of scientific concepts, the use of tactics designed to scare people into making radical and costly changes to the way they live is particularly distasteful. Exaggeration of scientific hypotheses, the misunder-standing and selective editing, or deliberate distortion, of findings to support preconceptions amounts to a misuse of information. And, in the end, it will be the scientific enterprise that is most likely to suffer through loss of credibility if it discovered that the claims were bogus.

A fear of climate change is but one aspect of the fear of the future and its scientific basis is much weaker than campaigners pretend. It may be that further investigation will reveal a genuine problem that will need to be addressed, but this does not mean that we must reject the feasibility of a future that is pleasanter than the past. It is by solving problems that we progress and by exaggerating them we scare ourselves into inertia.

18

Nuclear Power –
A Faustian Bargain?

Nothing alarms people quite so much as the merest suggestion that a nuclear power plant might be built in the neighbourhood. Their fear is not based on an assessment of the actual likelihood that the plant will cause them any injury, but neither is it altogether irrational. Risk assessment is now a thoroughly reputable discipline, in which historical evidence of past injuries and the size of the population likely to be affected by any incident capable of causing injury are used to calculate the statistical probability and severity of harm to each individual. All such calculations show unequivocally that, Chernobyl notwithstanding, nuclear power plants are very safe indeed. The chance of being injured through living close to one is far smaller than that of being harmed in almost any other way. It dwindles almost to nothing compared with the risk of injury through smoking cigarettes, for example, and is a great deal less than the risk of being injured in a road accident or murdered. Whether or not a risk is acceptable depends on the balance between the risk and the advantage that would be gained by running it. Since in this case the risk is slight and the advantage great, you might think nothing remained to discuss.

Obviously, it is not so. The technique is essentially actuarial, but to those who fear the nuclear plant or any of a variety of installations and related activities, that is irrelevant. We are not computers and we do not evaluate our lives in actuarial terms. What matters just as much as the actuarial risk is the way that risk is perceived, and in most cases the two conflict. This, too, is the subject of much study. We perceive a risk as much greater if it derives from something over which we believe we have no control. Everyone should know by now that cigarette smoking, driving fast cars and heavy drinking are dangerous activities, but they are activities over which we have control. We can choose whether or not to participate in them. Similarly, we can choose not to walk alone at night in a neighbourhood where there is a chance we might be attacked. We cannot control the activities of an industrial plant. If there is an accident, we will be its helpless victims. A further consequence of this fear is the feeling that local ties are weakened, that the community as a whole is losing control and that this may lead to a loosening of the social fabric itself.[1]

The feeling of helplessness also increases the seriousness of the risk as we perceive it. We are influenced by the confidence that we place in the estimate of risk presented to us. Some dangers are better known than others. This can lead to disagreements among scientists and, in any case, all inferences drawn from scientific research are somewhat tentative because of the impossibility of finding positive proofs. Scientific 'facts' are to some extent probabilistic constructs.[2] Our private perception of risk can seem to us a more reliable guide than a heavily qualified scientific assessment.

Conflict can arise even over developments that people accept in principle, once it becomes necessary to find a site location. So strong are local objections that in the United States it has now become almost impossible to build unpopular facilities anywhere[3]. In most countries the response has been to involve people more directly in decisions that affect them, especially by providing them with full scientific and technical explanations of the issues and risks involved. Success is by no means assured, however, not least because of the widespread distrust of scientists, which is encouraged by opponents of innovation, often with the support of journalists. As long ago as 1974, a US survey reported that: 'over the last decade science writers often presented science from a more unfavourable viewpoint than in 1965. Over 80 per cent of the writers surveyed in this study admitted "writing articles in 1972 that could be interpreted by readers as unfavourable to science".'[4]

The anti-nuclear lobby

Public unease is understandable, but it can be manipulated. Some years ago it was proposed to conduct a geological survey close to a village not far from where I live. The survey was part of a nationwide investigation of geological structures that might be suitable for the construction of a long-term repository for nuclear waste, and it involved drilling test boreholes. Anti-nuclear campaigners arrived from far and wide to organize a protest that included blocking the entrance to the field in which the boreholes were to be drilled. The field was guarded day and night, people lay down in front of vehicles, and in the end the project was abandoned, despite assurances that the survey was necessary and that there was no intention of constructing the facility at this site or anywhere near to it. When literature was produced explaining what was to happen and the policy of the nuclear industry on waste disposal, the protestors seized and burned it before anyone had a chance to read it. At about the same time, the nuclear industry mounted a small exhibition in a nearby town. I saw campaigners picketing it and actively discouraging members of the public from entering. The clear aim of the anti-nuclear movement is to silence all opposition to their views and the aim has been very largely achieved; theirs are now the only voices heard.

When it began, the opposition to nuclear power was concerned not with accidents or emissions from power plants, but with the proliferation of weapons. The first reactors had been built, after all, to supply fissile materials for the manufacture of weapons; their use for power generation came later. Amory Lovins, an early critic of the civil nuclear industry, wrote in 1977: 'the pregnant seeds of proliferation that nuclear power sows fall not on a geopolitical desert but on ground already tilled by processes of rivalry and tension that nuclear power itself helps to activate. Because nuclear power facilities are ambiguous and have military potential, actual or planned possession of them may impel neighbouring countries . . . to seek similar potential, if only to "keep their options open"' (his italics).[5] To a generation who grew up under the threat of thermonuclear war and was denied, for security reasons, any simply written explanation of the scientific and technological principles of reactors and weapons manufacture, the risk seemed plausible.

It may well have seemed, as the protestors sometimes allege, that we had entered into a Faustian pact with the devil, mortgaging the lives of our descendants to acquire an apparently limitless source of power. Our very use of the word 'power' to describe a form of energy is itself revealing, suggesting as it does a means by which we may control the world around us. Combine this with the mystery surrounding the scientific principles underlying nuclear decay and the source of the power assumes an almost magical quality, fortifying the Faustian allusion with the hint of something unnatural and unwholesome. The allusion is absurd, of course, since the decay of uranium nuclei is a perfectly natural process, and capturing the energy that this releases is merely a matter of arranging enough of those nuclei in a suitable configuration. Nuclear power is no more unnatural than burning coal or oil.

Nor is the link with weapons as strong as it may seem. As Lovins admitted, reactors destroy fissile uranium rather than producing it, they do not use uranium in a form remotely suitable for making bombs, and all but a tiny proportion of the plutonium they produce undergoes fission inside the reactor itself. The remaining plutonium can be recovered through reprocessing, but that does not necessarily purify it to anything like weapons-grade quality. About 20 tonnes of plutonium are recovered every year from reactors throughout the world and under international law every last gram of it must be recorded in an inventory and accounted for. Cheating is virtually impossible, because reactors cannot easily be hidden and their plutonium production can be calculated. It might be possible to cause trouble by disseminating low-grade fissile material, though the persons doing so would be the first to suffer injury, but it would not be feasible to use it to construct an atomic bomb. As the war with Iraq demonstrated, the key to producing weapons lies less with reactors than with the much more difficult enrichment technologies. A real danger arises from the possible emergence of a black market in material removed from nuclear weapons that are destroyed, but this is not relevant to civil reactor programmes.

Uranium-235, the fuel for reactors and some bombs, occurs as 0.7 per cent of natural uranium. Some reactors are designed to use it at this concentration, others require it to be enriched to 3 per cent, by partially separating it from the more abundant isotope, uranium-238. Weapons require its enrichment to more than 90 per cent and achieving this degree of enrichment is technologically difficult and expensive. There is a genuine risk from the proliferation of weapons, but the civil nuclear industry contributes little to it and these days nuclear weapons are not the only ones that can cause appalling damage. Consider, for example, the fuel–air bomb and the wide variety of so-called 'smart' weapons.

Many doubts have been expressed over the need for producing plutonium in breeder reactors and recovering it through reprocessing, given the surplus of plutonium left over from disarmament at the end of the Cold War. The Japanese are planning to build reactors that will burn a 'mixed oxide' fuel, made from oxides of uranium and plutonium, using plutonium recovered from their own spent fuel at the THORP (Thermal Oxide Reprocessing Plant) in Britain, or they might be able to use suitably modified weapons-grade fuel. The French Government plans to convert its Superphénix breeder reactor into a power-generating 'furnace' that will use some of the plutonium surplus.[6] Breeder reactors greatly increase the efficiency with which reactor fuel is used, but the slowing down of building programmes for conventional reactors has averted a predicted shortage of uranium and for the time being rendered breeders economically unnecessary.

Fear of proliferation was always accompanied by a more general fear of radiation, also associated with weapons and often much exaggerated. 'The test explosions carried out at Bikini Atoll in 1946 have left a grim demonstration of the long-term effects of radiation in the legacy of bizarrely deformed plants and animals on the island,' wrote Gerald Foley in 1976.[7] In fact, the first explosion destroyed every living thing on the atoll and stripped away the soil. Twelve years after testing there ceased, scientists visiting the atoll found it being recolonized by perfectly ordinary plants and animals, but this kind of overstatement of the effects of radiation exposure leads quite naturally to the conclusion that no risk of any kind can be tolerated. Lovins wrote of fission technology that 'ultimately its safety is limited not by our care, ingenuity, dedication, or wealth . . . but by our inescapable human fallibility', implying that safety is unobtainable by definition.[8]

Radiation exposure, fact and fiction

Science fiction films have fuelled the widespread fear of the mutagenic effects of radiation, depicting a world populated by giant arthropods and hideously deformed subhumans. Foley plays on that fear, but it is based on a misconception. Certainly, ionizing radiation can cause alterations in nucleic acids, as can a variety of substances including the oxygen we breathe. These changes may prove cancerous, but they are also the power that drives evolution. If genetic material were invariably to replicate with total accuracy, new forms could not emerge. As Lewis Thomas has pointed out: 'The capacity to blunder slightly is the real marvel of DNA. Without this special attribute, we would still be anaerobic bacteria and there would be no music.'[9] The key word is 'slightly'. When the alteration is too great, the cell or the entire organism carrying it dies because it has lost the ability to assemble essential proteins. Grotesquely deformed mutants are biologically impossible. So are insects and spiders the size of horses and rats the size of elephants. For purely mechanical reasons, if you design an animal the size of an elephant it ends up looking very much like an elephant, and gargantuan arthropods would asphyxiate, because their bodies transport oxygen by diffusion rather than a circulatory system and diffusion is effective only over very short distances.

Nevertheless, exposure to large doses of radiation is harmful and dread of dire genetic consequences underlies the more overt fear of immediate injury or the enhanced risk of cancer years later. Were a reactor to fail catastrophically, radioactive substances might indeed be released. The film *The China Syndrome* explores what was once considered the most serious reactor accident conceivable, the commercial success of the film deriving from the coincidence of its release, in 1979, with the serious failure of a reactor at Three Mile Island, in Pennsylvania.

Inside the core of a conventional reactor, the decay of its fuel releases neutrons, with varying amounts of energy. High-energy, or 'fast' neutrons collide with other nuclei, but simply 'bounce' off them, producing no effect. For this reason, the fuel is surrounded by a moderator, a substance that slows the high-energy neutrons. When slow neutrons collide with a fissile nucleus they are captured, destabilizing the nucleus, which decays with the release of energy and more neutrons. This establishes a chain reaction. The nuclear decay produces heat, which passes through a cooling system to steam-driven turbines that generate electricity. It was suggested that should the cooling system fail and should that failure be accompanied by the simultaneous failure of the secondary, and sometimes tertiary, cooling systems designed to take over in such an event, the core would grow rapidly hotter until it melted its way through the basement of the building in which it was housed, its downward progression (fancifully in the direction of China, hence the name) ending at the water table, where the reaction of hot, intensely radioactive material with cold ground water would cause an immense explosion. In fact, the Three Mile Island accident did not lead to a meltdown of this kind, but the similarity was close enough.

The potential consequences of a meltdown were studied in great detail, both in theory and, in 1993, in practice when a partial meltdown was engineered deliberately in a French reactor. It was discovered that, depending on the type of reactor, a failure of the cooling system either causes the fission reaction to stop, so the core does not overheat, or allows the core to heat, but only slowly, allowing ample time to shut down the reactor. The 'China syndrome' is a myth and the Three Mile Island accident released almost no radioactive material.

No sooner had our fears been calmed, however, and the anti-nuclear voices muted, than on 26 April 1986, a Soviet RMBK reactor at Chernobyl in the Ukraine really did suffer a major catastrophe, with a huge release of radioactive material. It was by far the most serious reactor accident that has ever occurred and one that could not affect any other type of reactor. This is because the RMBK design uses a graphite moderator and is cooled by water carried in pipes through the core. Should water and graphite (carbon) come into contact with one another at high temperature and pressure, they can react to form 'water gas', a highly explosive mixture of hydrogen and carbon monoxide. To minimize the risk of this, the RMBK core is contained within an atmosphere of nitrogen, but the design has long been considered inherently unsafe by engineers in other countries. Many designs use graphite as a moderator and many use water as a coolant, but no other design combines the two. The British Magnox reactors, for example, use graphite as a moderator and carbon dioxide as a coolant.

In the years that followed the Chernobyl accident, the health of people in the most seriously contaminated areas was monitored closely by international teams of scientists and medical personnel and there are detailed medical records for well over one million individuals. Not surprisingly, there were vast numbers of medical complaints. Most of the victims attributed their illnesses to radiation, but doctors found them due mainly to the psychological and social upheavals following the accident and to nutritional disorders arising from fear of eating contaminated food, but every complaint was examined carefully, even toothache. This afflicted many people, apparently, because their restricted diet contained truly amazing quantities of candies and other foods made from or rich in sugar. The studies were complicated by the poor health of the population prior to the accident and the inadequate medical facilities: some 30 per cent of hospitals in the affected area had no running water.

Some early reports were dramatic. In February 1989, for example, there was said to have been a large increase in the number of livestock born deformed on a collective farm 50 kilometres south-west of Chernobyl, outside the 30-km exclusion zone. In 1987, 37 piglets and 27 calves were born with severe deformities, and in the first months of 1988, 41 piglets and 35 calves, compared with three deformed piglets and no deformed calves recorded in the five years preceding the accident[10]. In fact, the radiation dose required to produce such a large effect would have been sufficient to kill humans exposed to it, but the health of the farm workers seemed to have been unaffected. Nor were neighbouring farms affected; the case was isolated, and the report was soon forgotten. Some claims may have arisen from statistical misunderstandings. Of the 600,000 people engaged in the clean-up operation it was said that within four years 7,000 to 10,000 had died. The figure sounds startling, but in western Europe, in a population of 600,000 healthy people aged between 35 and 40, about 7,000 would be expected to die in any four-year period. If the Chernobyl figure is true, not all of the deaths can be attributed to the accident. This is not to say there were no ill-effects from the accident. A large area of pine forest was killed and many other trees were injured, broadleaved as well as coniferous species. There was a decrease in populations of soil fauna near the plant and rodent populations also fell, though they recovered quickly once the humans had been evacuated. Populations of other species also benefited from the cessation of hunting.[11]

Seven years after Chernobyl, the World Health Organization reported that 'so far, no increase in the incidence of the radiation-triggered cases of leukaemia has yet been found.'[12] Confirmation that there had been no increase in childhood leukaemias between 1979 and 1985, before the accident, and 1986 and 1991, following it, was provided in October 1993, by doctors in Minsk, Belarus.[13] There was an increase in the number of cases of thyroid cancer reported among children in Belarus. The incidence rose from two to five cases a year between 1986 and 1988, to 55 cases in 1991 and 67 in 1992, with a total of 168 cases diagnosed between 1986 and 1993 compared with seven cases in the seven years preceding the accident. More than half of the cases were reported in one area, however. Other thyroid disorders not linked to radiation were also reported in Belarus but there was no increase in Russia or the Ukraine. So the rise is somewhat mysterious and the link with the accident far from established. It may be a statistical artefact, due to the increased attention paid to a complaint that was formerly under-reported.

The International Chernobyl Project, an earlier study conducted under the auspices of seven international organizations, compared the health of people living in contaminated areas with those living outside them, but did not examine the 600,000 'liquidators' involved in the immediate clean-up operation. This would have been extremely difficult and costly, because by then the people concerned had returned home and were scattered throughout the country. The Project found many illnesses due to psychological stress, but 'there were no health disorders that could be directly linked to radiation exposure. There were no indications of an increase in the incidence of leukaemia or solid cancers'.[14]

There was disagreement, however, over abnormalities in human pregnancies, some leading to stillbirths. The international team reported no such effect, but at the four-day meeting held in Vienna in 1992 to present its findings, the vice-chairman of the Ukrainian Academy of Sciences, Viktor Bar'yakhtar, said these had been recorded at a statistically significant level. Studies of the survivors of Hiroshima and Nagasaki, on the other hand, where there has been no significant increase in genetic effects, suggest it is unlikely that such effects would result from the doses to which the population was exposed.[15] All children born within one year after the accident continue to be monitored, especially for indications of brain damage, mental retardation and other neurological damage they might have suffered in utero. But by April 1993, the World Health Organization reported that the International Programme on the Health Effects of the Chernobyl Accident could find no evidence for this. Studies of the long-term effects are further complicated by the improvements in medical care introduced since the accident. In the Gomel district, for example, which was one of the most heavily contaminated areas, the death rate per thousand newborn babies was 16.3 in 1985, 13.4 in 1986 and 13.1 in 1987; in the Kiev region the figures for those years were, respectively, 15.5, 12.2 and 12.1.[16] This change must have been the result of better health care.

The fears of widespread illness and death were unfounded. True, in years to come there will be effects, principally an increase in the incidence of certain cancers, but these may be too few to measure. Among people living within the exclusion zone at the time of the accident, the pessimistic estimate is that there may be some 800 deaths from cancer induced by radiation, amounting to an increase of about 2.6 per cent over several decades. This might be detectable, but the figure may also be too high, because the improved medical services and screening now being provided may detect and cure a proportion of the cancers. In the former Soviet Union as a whole, the estimated increase in radiation-induced cancer deaths amounts to less than 0.03 per cent, and in the world as a whole to around 0.01 per cent. Such small changes will not be detectable against the natural fluctuations in cancer rates.[17] This may sound callous, for the figures refer to real people dying from real disease, but it should be seen in context. It does not imply that children and young people will be taken from their families, but in the main that elderly people will die sooner, by a few months or a year or two, than they would have done otherwise. This is not a happy situation, of course, but it does show that nuclear power, even from RMBK reactors, is less injurious than many people suppose.

The nuclear safety record

Nuclear plants have operated in Britain since 1950, not all of them generating power. In 1987 a report was published of a major study of the incidence of cancers found in their vicinity between 1959 and 1980. A summary of the report by a group of experienced epidemiologists stated, it produced 'strong evidence that there is no generalized increase in cancer mortality around nuclear installations in England and Wales either in young persons or in adults'.[18] Clusters of leukaemia cases in children have been found in the vicinity of nuclear plants, but no link has been found between them and radiation exposure and there are similar clusters in other areas far from any nuclear installation. The present view is that they may be caused by a virus.

This is also the verdict in the case of the children of men who were exposed to relatively high radiation doses while working at the Sellafield, Cumbria, plant prior to 1965, among whom there is a higher than normal incidence of leukaemia and non-Hodgkin's lymphomas. The evidence was studied closely, because, although the number of cases was small (seven leukaemias in people under 25 years of age living in Sellafield between 1955 and 1983), it clearly amounted to a cluster (the number was about ten times greater than might have been expected) and its implications were serious. The cluster occurred in only one village, yet 92 per cent of births of children to Sellafield employees took place outside Sellafield and their fathers received 93 per cent of the total radiation dose to all employees during the period in question. No similar link between paternal exposure and subsequent leukaemias could be found elsewhere in the world, including among the Hiroshima and Nagasaki survivors, whose fathers had received very much larger radiation doses. In the end, the study concluded that 'the association between paternal irradiation and leukaemia is largely or wholly a chance finding' and an alternative explanation should be sought for the small but real leukaemia clusters in young people near Sellafield.[19]

The final objection raised against nuclear power concerns the ultimate disposal of its wastes, often described as a problem to which no solution has been found. This, too, is wholly untrue, as is the claim that wastes must be consigned to secure storage for million of years. Depending on the composition of the waste, after up to one thousand years it will be no more radioactive than common uranium-bearing rocks, such as granite. There has been competition for the best long-term storage method, with several contenders, all of them effective, and the necessary choices have been made. The matter has never been urgent, because the volume of waste is very small, but there is no question of the industry accumulating difficulties that it is incapable of handling. Techniques for the decommissioning and dismantling of obsolete installations are also adequate, and the first decommissionings have already begun.

You might think the safety of civil nuclear installations had been established beyond reasonable doubt. The industry has existed for more than thirty years and in that time there has been only one really serious accidental release of radioactive materials, and even its effects were less severe than had been feared. Its safety record is much better than that of most other industries of comparable size. The evidence for its safety has been published, yet anti-nuclear feelings continue to thrive, fed most recently by disputes over the cost of nuclear power, which vary according to the way you compute them.

Almost everything in our environment is radioactive, including the food we eat and our own bodies. Burning of coal also releases radioactive substances. A power-station engineer at a coal-fired plant once put radiation-monitoring badges on the workers and found that radiation levels, though very low, were double those at a nearby nuclear plant. And that is not all that coal-burning produces. 'When Chernobyl's ruined reactor has become less radioactive than the soil in your back garden, when the Pharaohs' mighty pyramids have crumbled into sand, when our sun has become a Red Giant and boiled our oceans dry, the cadmium from your local coal station will be as deadly as on the day it left the smokestack'.[20]

Anti-nuclear fears concern only the supposed threat to human health, which makes the industry a curious choice of target for environmental campaigners. There is no evidence that the routine operation of nuclear power plants ever injured a single member of the public, nor the slightest suggestion that any non-human has been harmed, or could be. When protestors claim to have discovered radiation in plants or animals at marginally more than ordinary background levels, even they do not claim that the plants or animals show any sign of injury. Indeed, there are places in the world where natural radiation levels are very high indeed, but healthy plants grow there.[21] The choice is still more curious when made by those who simultaneously urge a reduction in emissions of carbon dioxide, since the 'natural' alternatives they offer hardly withstand scrutiny. In 1993, the British nuclear industry mentioned the possibility of building a new, twin-reactor plant with a 2.6 gigawatt capacity. Wind power is popular with environmentalists, though not with people living near wind farms. To match the output of that one nuclear power plant would require at least 14,000 wind generators of the type currently being installed (which are a little taller than Nelson's Column) and they would occupy around 14,000 hectares (35,000 acres) of land.

It is difficult to avoid the conclusion that anti-nuclear protestors are either ignorant or simply frightened of a future in which an industrialized society can thrive by the use of what was once an advanced technology but is now very well established, reliable and based on well understood science. Public ignorance is understandable, given the vigour with which informed debate is suppressed, but all the more worrying for that reason. The clear implication is that the anti-nuclear lobby is engaged in deliberately spreading misinformation. The effect, which is presumably intended, is to discredit scientific research and instil in people a fear of the future, causing them to reject any possibility of progress.

V

RESURRECTING TOMORROW

19

'Prediction is difficult, especially of the future'

Back in the 1940s, Saturday morning cinema shows were the highlight of the week for me and my friends. Filling every seat in the house, we saw Western serials, jungle adventures and cartoons. There was also a songsheet on the screen with a little ball that bounced along the words, and usually a space adventure serial. Occasionally some of those old sci-fi movies can be seen on late-night television. Once you have adjusted to the low-budget sets and costumes and the crudity of the plots and acting, the thing that strikes you most is how old-fashioned they all look. Set in the future, they give the clear impression that nothing dates faster than tomorrow. It is quite an achievement, when you think of it, to make futuristic movies look more out of date than those made half a century ago that were set in the present.

What jars is the technology. Spaceships of the 1940s are full of great metal levers, steering wheels and dials. They look as though they might be driven by steam. They are also huge. The crew move around a flight-deck that would make the bridge of a supertanker feel cramped, and they walk rather than float. Clearly, no one took much account of the inverse square law of gravity. With the infallible wisdom of hindsight, we can put the old movie-makers right on a few points, because we have seen pictures of the insides of real spaceships. They are cramped, people float in them and they are filled to bursting not with wheels, levers and dials, but with computer screens and illuminated control panels. They are not so much mechanical as electronic.

Prediction, time and the second law

Fifty years ago, no one predicted the existence of semiconductors or the effect that computers using them would have on the lives of everyone. The makers of low-budget movies for children were not alone in this. Prescient in many ways, even Jules Verne and H.G. Wells failed to predict modern electronics. Their futures were distinctly of the nineteenth- and early twentieth-century steam-powered variety.

If this is a criticism, it is intended as a very mild one, and it is easily rebutted. Writers of science fiction seldom attempt prediction as such. They write commentaries on the present, sometimes setting them in remote times or places for effect or amusement. It may also seem unfair to draw attention to the absence of phenomena that the authors could not possibly have known about. Half a century from now the world may well have changed yet again because of a future innovation, but there is no way that someone writing today about the future could allow for this. Any author who intends their work to be read as a genuine prediction is making a mistake.

Predictions are almost certain to be incorrect, and the best hope for those who insist on making them is to die before the passage of time exposes them to ridicule. It is said that in the early years of this century people worried about the increasing volume of horse-drawn traffic in London and calculated the date by which large parts of the city would be buried beneath manure. At that time few people believed motor transport would ever be more than a novelty or a plaything for the rich. The idea proved very persistent. Some years ago, a friend pointed out to me that many of the housing estates built in Britain in the 1920s and 1930s had provided no garages or parking facilities. The planners had not anticipated the rise in car ownership that was to occur a generation later.

In the 1940s, before he became the first president of Israel, Chaim Weizmann suggested growing starch crops for fermentation to produce ethanol, which could then be converted into butyl alcohol and propanone (acetone) to provide liquid fuel and a basis for several chemical industries. This, he believed, would free the world from its dependence on Middle Eastern oil. In the 1960s, oil was being used to produce food, by growing yeast on it.[1]

'To prophesy is extremely difficult,' says an old Chinese proverb, 'especially with respect to the future.' A prophet lives in a particular place at a particular time and under particular conditions and it is from these circumstances that all predictions must begin. They purport to show how the world will develop from a given starting point, but they do so by identifying what seem to be trends and then extrapolating them. Inevitably this produces a 'future' that, in fact, is the present with some aspects emphasized, and it is almost impossible to rid it of bias. If I find a trend disturbing it will be very difficult for me to avoid drawing attention to it as a warning, while being strongly tempted to emphasize pleasanter trends. Furthermore, my personal likes and dislikes are conditioned by the culture in which I grew up and now live. The prospect of a future that I would find delightful might horrify someone else.

History generates demons to torment soothsayers and bring hours of innocent mirth to ordinary folk. Vehicles that can travel faster than 50 km/h (30 mph) without annihilating their passengers, a motor car for nearly every family, manned spacecraft, nuclear power, computers, and modern tele-communications are all demons that have popped up through the star trap in modern times to confound those who thought them irrelevant, improbable or downright impossible. I doubt whether many predictions made a couple of decades ago included the ending of the Cold War, the dismantling of the Berlin Wall, or the disintegration of the Soviet Union, not to mention the release of Nelson Mandela, his sharing the Nobel Peace Prize with former President F.W. De Klerk and the holding of free elections in South Africa.

Time, as suggested earlier, proceeds in a linear fashion. This makes progress feasible, but it is also what makes prediction impossible. If time were cyclical, the future might revisit the past and we could employ historians to foretell it, but it is not so, and for a simple, logical reason. To predict the future we must possess knowledge that will influence events in the future, but that knowledge will then no longer reside in the future, but in the present, so it will be to the present, not the future, that our subsequent statements refer. There is also a reason well known to physicists. The concept proposed by Christians that has shaped our history so profoundly has been confirmed by scientists.

Drop a cup onto a hard floor and it will shatter into many pieces. It never happens that the fragments of a broken cup reassemble themselves of their own accord. It is a question of probabilities and energy. There is an extremely large number of ways in which the fragments of crockery can be arranged, but only one way in which they can be arranged to form a cup, and it is a special, ordered way. Order can be achieved only by doing some work, as the potter did when the cup was made or as you will do if you collect all the pieces and glue them together again. When the cup shatters, out of all the possible patterns available the pieces adopt the one requiring the least energy to maintain. They undo the work of the potter and if you want your cup back again you must pay by expending energy.

Place a hot-water bottle in the bed and it will cool as the bed warms. It never happens that the bed grows colder and the bottle hotter until the water boils inside it. This, too, is a matter of energy and order and it was first explained in 1850 by the German physicist Rudolf Clausius (1822–88), who wrote that 'heat does not spontaneously pass from a colder to a hotter body' (although it can be made to do so by means of a heat pump, which uses energy to 'pump energy uphill'). This is one of several ways of expressing the second law of thermodynamics, one implication of which is that the attainment of order necessitates the expenditure of energy. Warming the bed and making the cup require the application of energy, by heating the water or by the work of the potter. The outcome is increased order, in a bed warmer than the air in the bedroom or in a useful arrangement of pottery, and it is a one-way process. Leave them long enough and eventually the bottle, bed and room air will all be at the same temperature; sooner or later most cups are broken. Order is lost, disorder increases and, without further expenditure of energy, the order cannot be regained. This imposes a direction on time itself, sometimes called 'the arrow of time'.

Some physicists believe it might be possible to travel in time, avoiding the obvious paradoxes of the traveller into the past who kills his own mother before he was born and the traveller into the future who effects changes with implications for the present. The physics are complicated, however, and it will be some time before anyone tests a time machine, if such a machine is ever built. In any case, the directionality of linear time imposes a constraint that time travel would evade rather than remove. It ensures that the future will be different from the past. Its course will diverge from trends we can identify now, so that the further into the future we peer the more different it will be in ways we have no way of imagining.

Impossibility does not necessarily imply uselessness. Prediction is impossible, but we must plan for the future on the basis of reasonable assumptions, so we are careful. We do not predict, we forecast, and there is a difference. I forecast that one day I will be old and it would probably be sensible to plan for that contingency. But I should not take it for granted. For all I know my computer may electrocute me, I may lose a violent disagreement with a truck on my way to post a letter, my autonomous heart may decide it is bored with pumping, and this time next week I could be in the back of the big black limo on the slow ride to the crematorium.

Forecasts deal in possibilities, not inevitabilities, and this allows forecasters to explore opportunities. Years ago, Alvin Toffler advocated the establishment of 'imaginetic centres' in which people from many different walks of life could collaborate creatively to invent solutions to problems and devise futures they find pleasant. He quoted Christoph Bertram, of the Institute of Strategic

Studies, in London, who described their purpose as 'not so much to predict the future, but, by examining alternative futures, to show the choices open'.[2]

It might never happen

We are surrounded by doomsayers whose predictions overwhelm us, but what are presented as predictions are really no more than forecasts, and forecasts contain a large amount of opinion. If the forecasts are gloomy, then perhaps this says more about the forecasters than it does about the future we are likely to experience. At the risk of repeating what was said earlier, while humanity certainly faces problems that must be addressed, the unavoidably dismal future we are offered is a potent mixture of muddle and determinedly anti-scientific propaganda. The propagandists achieve their effect by selecting those issues most likely to cause distress and blaming scientists for them. This is intended to discredit in advance any scientific criticism of their analysis or proposal for straightforward solutions, and to present us with an array of difficulties so intractable as to destroy us.

If this sounds extreme, let me refer once again to Bryan Appleyard. After saying 'Everything is reduced by the idea of progress and change', he goes on: 'Science created the idea of constant forward movement and of the possibility of complete transformation. Its experimental methods and attitudes are implicit in the revolt of the young. And, of course, its ruthless rejection of the past is implicit in the impotence of the parents. For the truths of science do not require the wisdom of the past'.[3] Clearly unaware of the history of science as a discipline in its own right and of the deeply affectionate respect with which scientists are encouraged to regard their predecessors, Appleyard invents nonsense in order to dehumanize them and urges upon us an intellectual stasis that closes all our options. In his future, we will be able to do little more than punish our wayward children in the hope of instilling in them a proper observance of standards of thought and behaviour that he and others who think like him have declared 'traditional'.

We cannot allow those who dread rationality to stop us thinking or give any encouragement to those who would impose on us an authoritarian social order based on nostalgic sentimentality. The idea of progress should be welcomed in the belief that just as the present is much better than the past for many people, so the advantages it has to offer may benefit those presently less fortunate and the improvement may continue. As we find the courage to rediscover the future, we must allow that scientists will contribute greatly to its realization. The business of scientists is the acquisition of knowledge essential to our understanding of ourselves and the universe we inhabit. That is the activity so bitterly opposed by the anti-scientists, but it threatens no one. That is the activity the antiscientists so bitterly oppose, but it threatens no one and scientists are more than willing to participate in resolving such ethical implications as it raises. Suppose, then, that we classify all the gloomy prognostications as opinions, neither more nor less valid than an opinion anyone might hold. This permits us to admit alternative opinions and to give them equal recognition. That is what I propose to do in the next few chapters, by exploring just three examples.

Think of the adventures to be found in the exploration of space and, one day, the colonization of the Moon and Mars. Imagine how we might respond to the discovery that we are not the only intelligent beings in the universe. Consider the benefits to be gained from our rapidly expanding understanding of genetics. These topics cannot be discussed without some exploration of ethical issues, which I can outline but not resolve, and the nature of the scientific enterprise itself, which I can attempt to define. If we are to build for our children one of the possible futures that are attractive and available to us, we must bridge the gulf separating what have been called 'the two cultures', a task that requires a willingness on the part of all of us to learn more about what scientists do and what their findings mean.

20

Migrants from Earth

In August 1992, signals from the two spacecraft *Voyager* 1 and 2 included intense, low-frequency radio emissions they had received from outside the solar system. At first, scientists at the Jet Propulsion Laboratory were puzzled, then they realized what was happening. A few months earlier, the Sun had emitted a cloud of fast-moving, charged particles. These had reached the edge of the solar system where they encountered the charged particles that comprise the cold gas of interstellar space. The reaction between the two emitted the radio waves detected by the spacecraft.[1] The boundary where the solar system meets interstellar space is called the heliopause and the two spacecraft were crossing it. Launched in 1977, *Voyager* 2 on 20 August and *Voyager* 1 on 5 September, the two craft have now left the solar system.

The business of designing, building, launching, controlling and processing data from space vehicles is extremely technical. Once interpreted, however, their messages cannot fail to thrill anyone capable of experiencing wonder and delight. They tell of other worlds and regions of the universe we are visiting for the first time, albeit by proxy. Romance aside, the probes serve a very useful purpose: they provide information that is highly pertinent to our understanding of the origin and early history of our own world and may help us identify, in advance, threats to its welfare. They contribute to the survival not only of humans, of course, but of all earthly species. This knowledge enhances rather than detracts from the romance, transforming such voyages into purposeful explorations from which we can all benefit.

Joseph Campbell recorded his feelings the night Neil Armstrong first stepped on to the surface of the Moon, quoting what he described as the only adequate comment, by the Italian poet Giuseppe Ungaretti in the magazine *Epoca*: 'We see a photo of this white-haired old gentleman pointing in rapture to his television screen,' says Campbell, 'and in the caption beneath are his thrilling words: *Questa è una notte diversa da ogni altra notte del mondo*. For indeed that *was* "a different night from all other nights of the world"!'[2]

Our ancestors began exploring the Earth as soon as they had mastered the building of ships robust enough to carry them across the seas. Exploration in the west began with the Egyptians and as the technology improved the voyages grew longer. Herodotus describes how the Phoenicians sailed south from the Red Sea, completely around the coast of Africa, and three years later returned to Egypt.[3] There is a sense, of course, in which our explorations began much earlier, with those first migrations from Africa, where our species evolved, into Europe and Asia. You might say that the need to explore is 'hard-wired' into us and for a very good evolutionary reason: those species thrive best that are most adept at identifying possibilities and developing ecological niches for themselves, and humans are very good at it indeed.

From starting on foot, we invented ships, eventually ships with a keel that allowed them to be steered more efficiently and sails to power them, and the wheel to facilitate travel over land. As we went, we invented maps and the magnetic compass. A great deal of our inventiveness has been directed toward improving the means of transport available to us and recently the process has accelerated. My father never flew in an aeroplane and when I was a boy everyone looked up when one flew overhead. Later, I was trained to fly jet aircraft. Now we can send people into space and have visited the Moon. In September 1993, the Delta Clipper-Experimental rocket took off and landed, for the first time, at the White Sands Missile Range in New Mexico. Delta Clippers will be built for half the cost of a Boeing 747. They will reach orbit without using expendable lower stages or boosters and land again vertically, using their own engines to slow their descent. Once landed, a ground crew of three can prepare them for their next flight within 48 hours. Within a few years, Deltas may be providing scheduled passenger and freight services into space.

As we all know, the consequences of our explorations were not always benign. European explorers and the colonizers who followed caused considerable harm to many of the peoples they encountered and relationships were exploitative. Greed and zealotry were powerful motivating forces. Accounts of the early voyagers are replete with descriptions of battles fought, indigenous inhabitants slaughtered, slaves captured and wealth plundered. We may hope that the centuries have brought us wisdom. Unfortunately, now that only a few small, very remote regions of the inhabited world remain unknown, it is too late for us to do more than attempt some redress for past injuries.

The case for exploring space

Some people oppose all exploration of space, and most especially exploration with a view to the colonization of other worlds. They suppose us still to be motivated by greed and capable of violence. They may be right, but they are also wrong. Space is uninhabited and so are those other worlds presently accessible to us. They contain no living organisms we might injure and, just to make doubly sure, all planetary missions include the most thorough search for the slightest hint of life, allowing for the fact that organisms evolved elsewhere than on Earth may be quite unlike those with which we are familiar, and possibly difficult to recognize. Should life be found, scientists will insist that it be protected, if for no other reason than that it will afford unique opportunities to study the way organisms have evolved under different conditions and in complete isolation from Earth.

Absolute certainty about whether such life exists is of course unattainable, and the attempt to achieve it would be very costly and disruptive to the planet itself, but in reality the task may be less difficult than it seems. The Gaia hypothesis (see chapter 10) grew from the realization that the presence of living organisms, of any kind whatsoever, can be detected through their perturbation of their chemical environment. The technique was tested in December 1990, when the Galileo spacecraft flew past Earth at a closest approach of 960 km and data from its instruments allowed scientists to infer the existence of life on our planet, and to conclude from radio transmissions that the Earth supported a technological civilization.[4] Should intelligent beings be encountered on another world we will face the most interesting dilemma our species has ever had to resolve (see the next chapter).

It is also said that we have no right to despoil the environment of other worlds. It is difficult to attribute intrinsic value to an 'environment' that harbours no living organism and that no one has ever seen. We ascribe rights to all humans, some people would extend these to non-human animals and, in advocating the conservation of plants and habitats, we ascribe to them a right to be protected. But those who would ascribe rights to remote and inanimate rocks may be carrying a good thing a little too far. Instead, we may delight in the thought that human visitors may one day stand in a Martian landscape, gazing upon a distant prospect of Mons Olympica, at 24 km (almost 15 miles) possibly the highest mountain in the solar system, or walking in Valles Marineres, a system of canyons, about 4,000 km (2,485 miles) long and in places 700 km (435 miles) wide and 6 km (nearly 4 miles) deep, in which the Grand Canyon could be lost without trace.

When we commence our exploration of the planet, and I have not the slightest doubt that we will, the explorers will have to become self-sufficient in everything they require, because the distance is too great for them to be routinely supplied from the Earth–Moon system. This means the exploration team will be established as a colony, most of whose members will be employed in growing food, liberating oxygen from Martian rocks and processing Martian ores into metals and other materials. There can be little doubt that in the case of Mars, but only of Mars, they will also be engaged in a planned modification of the climate and atmosphere that will make it possible for the descendants of micro-organisms and plants brought from Earth to grow in the open. Mars will be 'terraformed'. This will make it more Earth-like, but it will never become a full replica of Earth. Mars will always be Mars and, before long, Martian people will be born and raised there.

James Lovelock proposed a way this terraforming might be accomplished and he and I wrote a semi-fictional book about it.[5] Briefly, our scheme involved injecting chlorofluorocarbons (CFCs) into the atmosphere to act as greenhouse gases, at the same time distributing dark-coloured lichens, obtained from Antarctica, in the hope that they might survive and spread to darken the surface colour and so absorb more solar radiation. As the atmospheric temperature rose, solid carbon dioxide from the polar regions would vaporize, increasing the atmospheric pressure and contributing to the greenhouse effect. This, we hoped, would raise the temperature sufficiently to begin melting some of the permafrost, which underlies large areas of Mars, causing water vapour to sublime (the pressure of the Martian atmosphere is too low for water to exist in the liquid phase and it moves directly between solid and gas). Water vapour is also a greenhouse gas. Colonists would start arriving early in this process and eventually they would be able to work in the open without wearing protective clothing, though they would have to carry breathing apparatus. The idea aroused considerable scientific interest and was studied in detail. The conclusion was that it might very well work, though Lovelock and I had greatly under-estimated the quantity of CFCs that would be needed and the time it would take to complete the terraforming.[6]

A team of scientists at the Obayashi Corporation, in Japan, have approached the same task in a different way. They would begin with an orbiting space station, then build lunar bases to provide experience of construction techniques in an extraterrestrial environment and for the production of oxygen, and use unmanned missions to explore Mars before the arrival of humans. They estimate that people could begin to arrive in 2020 and a colony of 150 could be established by 2057. After that, expansion would be rapid and the Martian population might reach 50,000 by the end of the next century.[7]

It makes sense to prepare major space programmes on the Moon rather than on Earth. Quite apart from the experience of working away from Earth that lunar conditions would provide, the low gravity and resultant low escape velocity would make it much cheaper to launch heavy payloads from the Moon. The interplanetary spaceships might then be built in orbit from materials ferried from the Moon. The Obayashi Corporation is active in this area, too, and planning more than factories. If the plans are realized, by 2050 its lunar city, built in a crater, would have 10,000 residents and large numbers of tourists, with leisure centres in which people could have fun exercising and flying hang gliders and human-powered aircraft in low gravity.[8]

Humanoid robots and starships

Scientists at Tokyo University are developing the first robot capable of repairing itself. Admittedly, the machine is a photocopier, but once established the principle is capable of extension and robots are becoming extremely versatile.[9] Mark Tilden, of the University of Waterloo, Ontario, builds robots that seek light to charge their batteries, and others that clean windows and the floor, none of which costs more than a few dollars to make.[10] Raj Reddy, of the Carnegie Mellon University, designs robots that drive cars and the first 'driverless' van (there is a real driver in case of malfunction) is already on the streets of Pittsburgh. The prototypes can drive safely at 88 km/h (55 mph) and one of them is able to find a parking place and manoeuvre into it.[11] A team at Waseda University, Tokyo, led by Ichiro Kato, is designing the first 'humanoid', a prototype robot descendants of which will be personal servants and assistants, especially to elderly or disabled people, and they will look much like the humanoid robots in movies. 'If it doesn't walk and act like a human, it isn't a robot – it's merely an automation,' Kato has been quoted as saying.[12] Away from Earth, robots have obvious utility, which increases with every increment in their capabilities. Machines that can move freely and safely in environments hostile for humans, providing their own power from sunlight and able to cope with minor breakdowns without human help, will be invaluable assets to teams of explorers and colonizers.

Mars and the Moon are the only bodies in the solar system we are likely to colonize, at least in the near future. Venus and Mercury are much more inhospitable and the satellites of the outer planets are very remote by our present standards and extremely cold, though scientifically fascinating. We are not restricted to them, however, nor eternally confined to the solar system. Schemes have existed for many years for the construction of huge vehicles at various locations within the solar system that would accommodate large human populations. They would be expensive, but one day it may be feasible to build such artificial worlds, though possibly more difficult to persuade people to live on them. One day, more certainly, we will set about building the first starship carrying people and some time later it will depart the solar system, following the two *Voyagers* into interstellar space.

Star travellers who embark from Earth will never reach a destination. They will inhabit a complete and enclosed world, but one that offers alluring benefits. Of necessity it will have its own system of government and eventually will develop its own unique culture. There the voyagers will live, and their children and grandchildren after them. When at last the ship reaches another star system, such planets as it may have will be explored by humans several generations removed from those who first embarked. You may be sure those pioneers will embark, moved by the questing spirit that carried their forebears to the remotest corners of the Earth and their more recent ancestors to the Moon and Mars, places that in their day were dangerous and inhospitable. Exploring is what humans do.

This urge to travel, to colonize planets and remoter regions of space, is often misunderstood. Some people, as has been said before, regard it merely as a means of obtaining material resources they believe will one day be scarce on our own planet. This may happen, but it is not the most important reason for exploration, though people living far away will have no choice but to exploit local resources for their survival and comfort. Increasing dependence on extraterrestrial resources may offer certain advantages, however. Freeman Dyson has suggested that: 'Earth may be treasured and preserved as a residential parkland, or as a wilderness area, while large-scale mining and manufacturing operations are banished to the Moon and the asteroids'.[13] This raises the more serious doubt that we will be tempted to move in order to escape from our difficulties and inevitably will find them within ourselves and travelling with us. Problems we identify here on Earth must be solved here on Earth, for they are of us, not of the planet.

We will explore because we are explorers, holding within us an enduring dream of the unknown. In a discussion with NASA Administrator Dan Goldin, Carl Sagan expressed this ideal and the spiritual justification for space research generally. 'Humans for 99 per cent of our history were hunter-gatherers,' he said. 'We wandered. We followed the game. Exploration is built into us. And just at the moment when the planet is all explored, save perhaps for under the ocean, the planets open up as a goal for exploration. Then there are the deep questions that each society, one way or another, asks – the origin of life; the origin of our planet; the origin, nature and fate of the universe. I think you'd have to be made out of wood not to wonder, just a little, about these questions . . . So it is reasonable for us who can, for the first time, actually find out some of the answers to make this investment.'[14]

Our exploration will carry us toward the future and we will travel hopefully, because journeys imply hope and the unknown is pregnant with potential. When humans resume the systematic exploration of the Moon and commence the exploration of Mars, we will be entitled to see them as symbols of that hope. When the first humans leave the Earth to take up residence elsewhere, the fact of their departure will show us how much freedom we enjoy to make the future as we would most like it to be.

21

Searching for Intelligent Extraterrestrials

We will never know with certainty that we are the only intelligent beings in the universe. It is impossible to prove a negative proposition and the absence of evidence cannot be taken as evidence of absence. All we can say is, at present, we know of no civilization other than our own. But if we were to discover unambiguous signs of such a civilization, that would constitute definite proof that we are not alone. The discovery would do more than change the course of our history; it would call into question many of our ideas about ourselves, especially religious ideas, and it might have a powerfully unifying effect.

People tend to react emotionally to the idea that other civilizations may exist and to attempts made to locate them. Some are so keen for us to contact 'aliens' that they grow impatient with the cautious procedure of scientific investigation. Others, consciously or unconsciously insisting on the uniqueness of our species and its privileged status, react angrily to the suggestion, dismiss it out of hand and fiercely oppose any idea of funding such an investigation. Between these two extremes are those who see, in the absence of any overt contact, a possible warning to us. Civilizations that have reached the level of technological competence required for contacting outsiders may already be on the verge of their own self-destruction – a consequence of that same dependence on technological development.

In other matters, fears and doubts can usually be stilled and disagreements resolved either by suppressing them or by providing the information on which rational judgements can be based. But, in this case, it would seem that we have little choice. Thoughts cannot be suppressed and our natural curiosity demands that we try to find definite answers. The search, which began some years ago, also has a more compelling purpose. We ourselves are now clearly visible to any beings who may be scanning the sky in search of civilizations comparable to their own. Since about the 1940s, the intensity of

216

electromagnetic emission in the radio wavelengths from our planet has increased about one millionfold, making Earth the second most powerful radio source in the solar system, after the Sun itself, and its brightness continues to increase. The day may well come, and soon, when we outshine the Sun at these wavelengths.[1]

Who's watching who?

The attention of any distant radio astronomer observing our star at these wavelengths will be drawn at once to so bright a planet and a study of the emissions will reveal certain very significant aspects of them. They occupy a narrow band, their central frequencies remain constant over prolonged periods, whereas the frequencies of most natural emissions drift over long periods, and all the signals contain a regular pulse which modulates their amplitude, a pulse lacked by natural emissions.[2] They are clearly signals conveying information. The combination of rapidly increasing brightness at radio wavelengths and the character of the emissions would lead any observer to infer the presence of a civilization which had recently mastered radio technology. Already, our presence may be known. If so, it would be wise to try to find out the possible consequences of having attracted attention to ourselves and to search for evidence of other civilizations. These would necessarily be technologically advanced, for it is only their technologies that would allow us to detect them, and they might be much more advanced than we are.

Once scientists believe they have found evidence of an extraterrestrial civilization, their first task will be to inform certain of their colleagues who will check the evidence and seek to confirm or refute it. If they, too, believe the signs are authentic, the secretary general of the United Nations will be notified. Only then will a public announcement be made, with the authority of the United Nations. From the outset, therefore, the discovery will be seen as affecting everyone on Earth, and ethnic, cultural, religious, political or national differences will suddenly seem very parochial. Peoples have united in the past, for trade, locally to oppose another grouping, or within an empire, but the alliances have usually been temporary and fragile. Knowledge of the existence of another civilization in the universe will demonstrate our common humanity and the evolutionary links that join all species living on Earth. We will have a new identity, as 'terrans', to be contrasted with the identity of those beings not of the Earth. We will be forced to recognize, for example, that biologically and historically we humans share more in

217

common with the humblest bacterium than we are likely to share with any extraterrestrial. The 'brotherhood of man' will be seen as a simple statement of fact rather than an ideal.

If some sort of confrontation is implied here, that is how it may be. We know from our own history that when two cultures meet, one technologically more powerful than the other, the weaker culture is likely to be overwhelmed. In any encounter with an extraterrestrial civilization, therefore, we should be aware of the risk we might be taking and remember that it may be much more advanced than our own. A mere ten thousand or so years have passed since humans began domesticating livestock and cultivating plants and no more than a few centuries since we began manufacturing goods on an industrial scale. It is perfectly feasible that another civilization may have achieved our present level of expertise a thousand or even a million years ago. Even if we assume sufficient common experience to permit communication between beings evolved in different regions of the universe, the cultural gap between us could be far wider than between, say, the European colonists of the New World and the indigenous peoples they met there.

So far, it seems that no attempt has been made to contact us and this may be a hopeful sign that others, if others there be, are familiar with the dangers. Deliberate contact might well be interpreted as a sign of hostility. Some people believe contact, or even visits, may have been made in the distant past,[3] but any beings interested in studying Earth closely might be expected to use the Moon as a base for their excursions, probably locating their installations on the dark side, where they would always be hidden from direct observation by Earth-based observers. The dark side of the Moon has now been thoroughly surveyed, no sign of such installations has been seen and so probably they have never existed and we have not been visited. This leaves open the remote possibility that our presence is known to observers who behave as we might wish our ancestors had behaved on Earth. Alert to the dangers arising from intercultural contact, they seek to observe us, to learn about our planet, but to do so without interfering with us in any way. They remain hidden from us and so we are unaware of them. It is extremely improbable, but not quite impossible.

The ongoing quest

Radio emissions apart, a spacecraft passing close to Earth would have no difficulty detecting the presence of life, but might well fail to notice the existence of a technological civilization. We know this from the experience of the Galileo probe, which was placed into orbits that carried it once past Venus and twice past Earth, using the gravitational attraction of the two planets to accelerate it into the orbit that carries it to Jupiter, where it will arrive in December 1995. On its closest approach to Earth, its most detailed photographs of Earth, with a resolution of a few kilometres, provided no evidence at all of the presence of intelligent beings. True, these photographs were of Australia, which is very sparsely populated, and of Antarctica, but any craft observing any planet might equally miss densely populated regions, if such regions exist. The Galileo data suggested that Earth does not support a 'Type B' civilization.[4] Ours is classified as a Type A civilization, one that makes only modest changes to the surface of its planet. A Type B civilization alters most of the surface, so it is much easier to detect.

Obviously, modification of the landscape can be detected, if at all, only by close observation. So great are the distances involved in travelling even to the nearest star, and much further to the outer reaches of our own galaxy, that direct contact of this kind is less likely than the passing of radio messages. For this purpose a different classification of civilizations is used, based on the energy the civilization uses as a proportion of the total energy it receives from its star. In this scheme, ours is a Type 0 civilization because we use energy equivalent to about one ten-thousandth of the energy that we receive from the Sun. A Type I civilization would use energy equal to the whole of the energy it receives from its star, or about 10,000 times more than we use, and a Type II civilization would use energy equal to the total power output of its star, or about 10 trillion (million million) times more than we do. The Russian astrophysicist Nikolai Kardashev, who devised this classification, also proposed a somewhat improbable Type III civilization, which uses energy equal to the total output of its galaxy.

SETI, the Search for ExtraTerrestrial Intelligence, began about thirty years ago. It was started in a haphazard way by enthusiasts who 'borrowed' radio telescopes during brief periods when they were not being used for other purposes. Today the search is more formal, on a much grander scale, and the name SETI is used only informally, to avoid any possible connotation of 'little green men'. The Megachannel ExtraTerrestrial Assay, or META, began scanning the sky over the northern hemisphere in 1985, with the help of funds from Steven Spielberg, and the High Resolution Microwave Survey, operated by NASA, began in October 1992. Bob Dixon of Ohio State University has been searching for radio signals full time for more than twenty years.[5] META has now been joined by META II, based in Argentina and scanning the sky of the southern hemisphere.

META, the most ambitious search, monitors more than eight million channels every 20 seconds. It concentrates especially but not exclusively on the 21 cm wavelength, which radio astronomers agree is the most hopeful. This is the wavelength at which neutral hydrogen emits radio waves naturally and, since hydrogen is by far the most abundant element in the universe, its characteristic wavelength will be familiar to radio astronomers anywhere. It is also a wavelength free from cosmic static. By international agreement, this wavelength is kept free for radio astronomers and no one on Earth transmits on it. Other civilizations might have a similar rule, so its second harmonic, at 10.5 cm, is also monitored. The computers recording the scan are programmed to identify particular characteristics. Interesting 'events' are then studied in detail to eliminate those that can be explained and a search is mounted for repetitions of any that remain inexplicable.

After eight years and five scans of the sky, META had accumulated 37 inexplicable 'events', but none of them has been repeated. In the words of Paul Horowitz, professor of physics at Harvard University and principal designer of META, 'as much as we might have hoped otherwise, the sky apparently is not teeming with civilizations beaming messages at us . . . Can it be that, among the 400 billion stars that comprise the Milky Way, not a single one harbours a highly advanced civilization?'[6] Horowitz and his colleague Carl Sagan have drawn preliminary conclusions from the META experience. They find that within 25 light-years of Earth there are no Type 0 civilizations (civilizations at a similar stage of development to our own) broadcasting uniformly in all directions at the wavelengths monitored, nor are there any in our galaxy beaming signals in our direction with a 300 m antenna, which is the size of that of the Arecibo radio telescope in Puerto Rico. There is no Type I civilization transmitting in all directions within 2,500 light-years of us, which means not one of some 10 million stars has a Type I civilization orbiting it, and there is no Type I civilization transmitting directly to us anywhere in the galaxy. There is no Type II civilization within 70 million light-years, meaning not only is there no such civilization within our own galaxy, there are none in neighbouring galaxies either.

Disappointing as they may be to those who hoped for more positive evidence, the META results do not prove that ours is the only species of intelligent beings. The search would have detected only certain types of signal transmitted at certain wavelengths at times when the receiving antenna was pointed in their direction. Far from being called off, the search is intensifying. META is to be joined by BETA, the Billion-channel ExtraTerrestrial Assay, which will cover many more wavelengths and will use two antennae rather than one, with a third to veto signals from Earth-based transmitters.

It may be that other civilizations exist, but communicate in other ways than by radio, for example by laser, or they might take precautions to conceal their transmissions. Earth shines brightly at radio wavelengths because our signals can escape through the ionosphere at night, but not during daylight. Perhaps they have found ways and reasons to prevent this 'leakage', to economize on power use, for example. More probably, in the view of the scientists engaged in the search, they do exist, they do transmit, but so far we have simply failed to detect them. However, until we identify an unambiguous signal, belief in the possibility of other civilizations is partly a matter of faith and partly of probabilities, but probabilistic calculations based on only one example, which is all we have, must be regarded with extreme caution.

We know that life began here very soon after the Earth formed, suggesting it might occur on any suitable planet. We know that intelligence, though it arose fortuitously, confers great survival value on its possessors, so its evolution is likely. Indeed, all animals on Earth are to some degree intelligent, at least in the sense of possessing sensory organs and a nervous system capable of interpreting signals from them; further advances are not so much qualitative as a matter of degree. We also know the galaxy contains many billions of stars, many of which are at about the same stage in their development as the Sun; some at least must have planets and some of those planets must be suitable for habitation, albeit by organisms very different from ourselves. Finding such a signal would, in the words of the physicist Philip Morrison, 'transform the origin of life from a miracle to a statistic'.[7]

The search will continue, because humans are impelled to seek answers and this is one of the most fundamental of all the questions we have ever asked. Should it bear fruit, at that moment the course of history will change. No longer will it be possible to think of the future as a mere continuation from the past. It will be qualitatively different because we will have been changed irrevocably by the new knowledge. It matters little whether the 'extraterrestrials' can offer us fresh ideas or splendid designs for new and better machines. In any case, a radio conversation conducted over a distance of many light-years would be ponderous in the extreme, with no real opportunity for a question-and-answer kind of dialogue. The importance will consist in the new sense it will provide of our own identity and unity. Perhaps merely knowing that 'extraterrestrials' exist, without our learning anything from them, would give humans the power to solve the disagree-

221

ments, conflicts and difficulties among themselves, and between themselves and their natural surroundings, which until then had bedevilled their every attempt to create a truly global, peaceful, just and secure culture. It would be difficult to think of anything that might be more exciting or produce more positive results.

22

Unravelling Genomes and Designing Species

The 1993 Nobel Prize for Chemistry was shared by two scientists who made major contributions to genetics. One of them, Kary B. Mullis, an American molecular biologist, invented the polymerase chain reaction, or PCR. This technique uses primers, consisting of short sequences of DNA that attach to either end of a particular section of DNA contained in a sample, and enzymes that cause this DNA to be replicated. Repeated PCR processing thus allows scientists to multiply a very small section of sample DNA into a large number of identical copies, which can then be cloned and read as sequences by laboratory techniques that are now routine.

There are many uses for PCR, including the study of archaeological material.[1] DNA has been obtained from human bones 11,000 years old, for example, and the book and movie *Jurassic Park* were based on the (scientifically dubious) idea that blood consumed by an insect which was then trapped in amber might contain DNA that could be recovered, using PCR. The technique is also being applied to plant remains, the eventual aim being to trace the evolutionary changes that have occurred in the course of the history of important crop plants, such as wheat.

Research into ancient DNA is now attracting much scientific interest, because the possibility it offers of tracing the evolutionary development of domesticated plants and animals, as well as of humans, provides archaeologists with a valuable new tool. It also allows scientists to study the evolutionary relationships among other species, including some of those that are now extinct. Our familiar farm animals and crops differ genetically from their wild relatives and their ancestors. Humans began genetic modifications the moment they selected certain individual animals and plants and bred from them, and farmers and breeders have continued the practice ever since.

Genetic engineering – an emotive issue

Jeremy Rifkind, the American anti-scientific campaigner whose Foundation on Economic Trends is bitterly opposed to all so-called 'genetic engineering', has launched a movement to boycott 'genetically engineered' foods. His Pure Foods Campaign has won support from 1,500 or more chefs who display in the windows of their restaurants a label stating 'We DO NOT Serve Genetically Engineered Foods'. Their principal target was a tomato containing an introduced gene to delay ripening, and so extend shelf life, developed by Calgene Inc., of California. To its credit, when the time came to market the tomato, Calgene labelled it boldly, charged a premium price for it, and it proved popular with consumers. I say 'to its credit' because our knowledge of the history of plant and animal domestication reveals the Rifkind campaign as misguided and the chefs as ludicrous. Henry I. Miller, of the Office of Biotechnology at the Food and Drug Administration, asked whether the chefs would boycott nectarines, which are genetically modified peaches, or the tangelo, part tangerine, part grapefruit and clearly transgenic. 'If a few sadly exploited chefs do make do with "non-genetically engineered" foods,' he wrote, 'their menu is likely to be a singular one, offering little more than fish, shellfish and wild game and berries.'[2]

Rifkind, originally an economist, declared his opposition to genetic engineering in 1977, in a document called *Who Should Play God?* His line of reasoning provides an interesting example of anti-scientific extremism. In 1983, in a book called *Algeny*, he wrote that Darwinism is dying, a fact that evolutionary biologists are entirely unaware of, and proposed that evolution is a process of 'mind enlarging its domain up the chain of species'. Each living thing, he wrote, is a 'pattern woven from the larger pattern' of mind that pervades the universe, so organisms should not be considered as assemblages of information to be rearranged according to human whim.[3] He used these obfuscations to form a coalition of American religious leaders anxious to 'resacralize' nature as Rifkind put it, to reassert the dominion of God over the mysteries of life.

Many religious people, of all faiths, resent what they think of as 'tampering with nature', but it is difficult to find any genuine basis for their unease beyond a simplistic technophobia. Religious communities often manage farms and are quite happy to exploit selectively bred species. Indeed, Gregor Mendel, whose experiments with peas laid the foundations for modern genetic research, was a monk who became abbot of his monastery. The objection seems to be not to genetic alteration itself but rather to the improved precision and efficiency of modern techniques as compared with the slower and more haphazard breeding methods that they replace, rather than to genetic alteration in itself.

There is no reason why some genetically altered products should not be clearly labelled as such. People whose religious views prohibit the eating of pig products should be warned of the presence of pig genes in products such as beef, and vegetarians might prefer not to eat products containing genetic material from animals. Labelling would address their anxieties, without suggesting that the products are anything but nutritionally sound, wholesome and perfectly safe.

The question of labelling was drawn to the attention of the British Government when a Scottish company applied for permission to sell lambs from its flock of genetically altered sheep. These animals contain a human gene for the production of a blood-clotting factor, called alpha-1-antitrypsin, that has to be supplied to haemophiliacs, who lack it. Hopefully, the substance will be contained in the ewes' milk, from which it can be extracted cheaply and in copious amounts. There is no question of the sheep being bred for meat production; they are far too valuable for that. Human genetic material has also been introduced into a Dutch bull, called Herman. All being well, this will cause his daughters to produce milk much more like human milk than the milk produced by other cows. Again, the benefits to human health are great and obvious. Since 1979, insulin, formerly extracted from the pancreases of pigs and cows, has been produced by genetically modified micro-organisms, and interferon, vaccines and human growth hormone are also produced in this way.[4]

Rational and irrational fears

Much has been made of the ethical issues involved in such manipulation, but it is not easy to see what they are, if they exist at all. Certainly we may question the exploitation of non-humans for the benefit of humans, but in this case the exploitation is much milder than, say, the raising of livestock for food. So far from being confined for their brief lives in intensive livestock units and then killed, these animals live well, they receive the best of attention and no one eats them. They are transgenics, since they contain genetic material from another species, in this case our own, but they are not aware of it and it causes them not the smallest inconvenience. They look, live and behave exactly like ordinary sheep and cattle, which is what they are, with this one very small modification. It is even harder to see what ethical issues are raised by the modification of bacteria or other micro-organisms, especially since we have been using *Sacharomyces cerevisiae* yeast to leaven bread and ferment beer for thousands of years, over which we have altered its genetic composition substantially.

Fears over the accidental escape and subsequent evolutionary success of modified organisms have a much more rational basis. Species taken from one part of the world and introduced in another have sometimes acquired pest status, usually because of the absence of their natural predators. The rabbit is the most familiar example, but by no means the only one. *Rhododendron ponticum* was introduced to Britain as an ornamental but has proved extremely invasive and now covers large tracts. *Eicchhornia crassipes*, the water hyacinth, was carried to the United States from Venezuela, also as an ornamental, by exhibitors at the Cotton Exposition in New Orleans in 1884. It is now a serious weed that clogs navigable waterways throughout the tropics and subtropics of both the Old and New Worlds. Two parent plants can produce 30 offspring by vegetative budding within 23 days and 1,200 within four months. The plant can also be harvested, however, and in Indonesia it is fed to pigs.[5]

These and many other examples serve as cautionary tales, but there is a difference. Genetically altered organisms are seldom exotic, in the biological sense of having been introduced to areas where they are not native, but are otherwise ordinary species differing in only minor respects from their well-established relatives. Nevertheless, their modified characters might confer an advantage that would allow them to acquire pest status, or they might interbreed with unaltered conspecifics to produce hybrids that became invasive. Not all modified organisms are designed for introduction into the environment, of course, and the great care taken to prevent their escape ensures that the risk of harm from them is small. Even greater care is taken where the micro-organisms being used could infect humans, perhaps to cause illness, and strict rules govern procedures in laboratories handling such species. That apart, it is very unlikely that micro-organisms developed for the commercial production of pharmaceuticals and adapted to the sheltered conditions of the laboratory or factory could survive long in the wild.

Some time ago an opportunity arose to test the consequences of releasing a genetically altered plant into the environment. The species concerned was a variety of oilseed rape (*Brassica napus oleifera*) that had been modified to make it tolerant of an antibiotic, kanamycin, and a herbicide, glufosinate. The study found 'no indication that genetic engineering . . . increased the invasive potential of oilseed rape. In those cases in which there were significant differences (such as seed survival on burial), transgenic lines were less invasive and less persistent than their conventional counterparts'.[6] Though entirely reasonable, the fears seem to have been unwarranted.

It is the idea of altering human genes that causes most alarm, not least because Nazi eugenics policies are such a recent memory (see chapter 12), though it may be worth recording that many churches supported eugenics in the early years of this century[7]. The issues seem clear enough. 'It may soon prove possible', said a *Daily Telegraph* leader at the time of a 1993 conference of geneticists, 'to "improve" the human breeding stock by genetic engineering, not only to eliminate hereditary diseases but also to cultivate intelligence, beauty or skill. But because it is possible, does that make it right?'[8] Apparently not, according to the article, but it is all a little more complex than the leader writer supposed.

In the first place, we have not yet reached anything like satisfactory definitions of 'intelligence', 'beauty', or 'skill'. Beauty is determined largely by fashion, skills by the requirements of particular historical episodes, and intelligence means different things to different people at different times. Even if the terms actually mean anything at all, there is no reason to suppose them to be wholly genetically determined. Skills can be taught, after all, 'intelligence' enhanced by education, and beauty can be achieved by the skilful use of cosmetics. In any case, beautiful, skilful and intelligent people tend to seek partners much like themselves in these respects, so if it is possible to select for such characters we have been doing it throughout history. Since this has not led to the emergence of demonstrably superior beings, the evidence would seem to suggest the impracticality of the enterprise.

It is nonsense, and harmful nonsense in that it diverts attention from a genuine ethical issue. Geneticists are identifying genes that predispose their carriers to certain illnesses and in years to come we may expect a genetic component of more diseases to be revealed. Already this is suggesting therapies, but the ethical question concerns access to details of the genetic composition of individuals and ways in which that information may be used and abused. There are fears, for example, that employers and insurance companies may discriminate against people who they believe are more susceptible than others to heritable ailments. Clearly this would amount to a gross invasion of privacy if not a major injustice, since an increased predisposition to a condition does not imply certainty that it will manifest itself.

The further fears are the eugenic ones: that the identification of potentially harmful genetic conditions in embryos will lead to the unwarranted termination of pregnancies; or that the techniques of genetic engineering will be harnessed to the elimination of those conditions from the human genome. These dangers are real and are being addressed, as they should be, before doctors and scientists are required to face them. At present they do not arise, because of insufficient knowledge, but that lack is certain to be remedied. Meanwhile, there are safeguards. In many countries any modification of heritable genetic material in humans is absolutely forbidden. Attitudes to abortion vary from one country to another, but in countries where permission must be sought or people considering abortion are counselled, approval would probably be denied if the only ground were a risk of transmitting illness to further progeny.

Meanwhile, cystic fibrosis, which is linked to a defective gene, has become the first heritable illness to be treated by gene therapy. First, a functional version of the gene is transferred into a virus. This is then transported to the affected tissue in the nose and lungs in the hope that the 'good' gene will replace a sufficient number of the defective genes to restore adequate respiratory function. Because the treatment is applied to somatic cells, the resulting genetic alteration cannot be transmitted to offspring.

Safeguards and benefits

Scientists recognize the ethical implications of genetic research, especially research into the human genome, and of the application of its results. Indeed, they have been to the forefront of efforts to bring these matters to the attention of an often indifferent public. Ethical committees that draw up rules and guidelines for researchers are composed mainly of scientists and were established several years ago. But it is essential that the more general debate be widened to include the public, as undoubtedly it will be. Decisions affecting society at large must be made by people drawn from many walks of life, but with one proviso. Those whom we appoint to reach judgement must be informed by scientists, not anti-scientists, so that judgement is based on what is real or may become possible rather than on the muddled fantasies of bigots who would prevent all research because they fear the creation of impossible monsters. There is a real risk that unreasonable fear could deny us very real benefits.

Modifying plants to make them resistant to pests and diseases will reduce dependence on pesticides, with clear environmental advantages. Improving the keeping qualities of plant products will also reduce the need to employ chemical preservatives, while supplying consumers with fresher food. More productive livestock will reduce the cost of food without causing more injury to the animals than they suffer in the ordinary course of husbandry and slaughter. The modification of non-humans to make them produce substances of value in human medicine does no harm to the non-humans but may make therapies available to vast numbers of people who at present are denied them because of their high cost or scarcity. New therapies will be developed, to treat defective genes directly, or to identify those associated with particular ailments, and supply victims with the proteins they are unable to produce for themselves. There will be benefits to conservation, as better understanding of the genetic composition of rare species is translated into breeding programmes.

Other benefits are less tangible, but no less real. Information resulting from genetic research will contribute greatly to our knowledge of ourselves, the ancestry of our species and our relationship with non-humans. Studies of ancient DNA will tell us more about the history and prehistory of our farming and of the plants and animals that live with us.

We are beginning to find answers to the most fundamental questions. The search for those answers is exciting and absorbing. Their discovery will augment our understanding of ourselves and our world. The more profound that understanding, the greater the chance that our descendants will avoid causing unintentional harm to themselves or to others and, therefore, that they will inhabit a future that is better than our present.

23

Science, the Continuing Quest

What is science? Although this book is about science, the word itself is rarely used, except in quotations from other authors. The reason is that, unlike 'scientist' and 'scientific', the word 'science' can be misleading. Literally, means 'knowledge', from *scientia*, the Latin word with the same meaning. However, according to the first definition given in the *Shorter Oxford English Dictionary* it is 'the state or fact of knowing; knowledge or cognizance of something specified or implied; also, knowledge (more or less extensive) as personal attribute'. Subsidiary definitions refer to knowledge acquired as result of study and the mastery of a skill (especially boxing!). Nowhere is the meaning confined to knowledge of a specified range of phenomena investigated according to particular methods by people we call 'scientists'. Thus, while physics or biology may indeed be described as 'sciences', according to the definition so might any other corpus of knowledge.

As the word is commonly used, by scientists, non-scientists and anti-scientists alike, it seems to imply a 'thing' towards which we can adopt an attitude. We can 'serve' science, support it, or oppose it, as though it were a 'cause'. This is to reify an abstract concept that has not been clearly defined and it generates nonsense. If 'science' means 'knowledge', albeit restricted to a special kind of knowledge acquired in prescribed ways, ordinary use of language suggests that its opposite is 'ignorance'. Doubtless anti-scientists do not think of themselves as preferring ignorance to knowledge, but if they believe there are non-scientific but equally valid means for accumulating knowledge on which they place greater reliance, they are still supporting 'science' in the strict sense. Support of any kind means little, however, for preferring knowledge to ignorance is akin to preferring good to evil. Our preference is predetermined by our understanding of the words themselves and they in turn are products of our culture. We are committed to our preference for the reason that knowledge of our surroundings and of our another makes us more secure. It would appear, therefore, that the acquisition of knowledge serves a very obvious biological purpose and requires no justification. However, this is only partly true. Not all knowledge is of equal value and not all knowledge is cumulative.

Knowledge, information and theory

Consider, for example, the work of two philosophers who lived several centuries apart but who are recognized as intellectual equals. Very likely their arguments and conclusions would differ, but the later philosopher can be said to draw from a larger body of accumulated knowledge than the one who lived earlier: he 'knows more'. That knowledge, recorded during the intervening centuries, must consist partly of arguments advanced and conclusions reached by other philosophers and partly of information about the universe. These are two different categories of knowledge. To show that the later philosopher knew more about the universe to which his arguments and conclusions related, we would refer not to the discussions of other philosophers, but to the larger volume of available information. Alternatively, consider whether the question 'Does the twentieth-century painter A know more about art than the sixteenth-century painter B?' makes sense. Excluding the history of artists and schools between the two centuries, all we can say is that while A may know of more pigments than B and may possess a more comprehensive knowledge of colour and the behaviour of light, this is information about the universe at large, not about art as such. (However, I accept that 'art' is as much a reification as 'science'. As Sir Ernst Gombrich points out in The Story of Art: 'There really is no such thing as Art. There are only artists.')[1]

Information about the universe at large does accumulate and this confers a quality of uniqueness on the knowledge derived from it. It is the type of information scientists collect and it makes sense to assert that a present-day scientist knows more about its subject, the universe, than any scientist who lived in the past. When a scientist embarks on a new research project, once the aim of the research has been clarified and the programme designed, it is standard procedure to make a comprehensive search through the published literature. Obviously, this helps avoid needless repetition of work already done, but its other purpose is to provide a solid ground of existing knowledge on which the new research can build. Research can be justified, and funded, only if it is likely to augment the corpus of knowledge.

231

Scientific etiquette enshrines this deep respect for history. The author of a paper reporting experiments or observations must always refer specifically to previously published work from which data or ideas have been absorbed. Furthermore, the importance of a piece of research is conventionally measured as the number of times the paper reporting it is cited by others. This convention has even acquired political overtones. Investigators comparing the amounts of significant research in different countries count the number of citations per country that appear in the published reports and use the results in discussing funding requirements with governments. Naturally, most citations are to recent work, but authors sometimes refer to publications from the early years of this century or even earlier.

Bryan Appleyard could not be more mistaken, therefore, in saying that 'the truths of science do not require the wisdom of the past', or in denouncing the concept of progress by identifying 'greatness' with the amount of information to which a person has access.[2] Admittedly, his claim that 'a computer scientist need never have heard of Newton' may be true, since the work of Newton is not of immediate relevance to computer research. Every physicist is familiar with Newton's work, however, and the computer scientist would be thought very ill-educated who had never heard of Charles Babbage (1792–1871), George Boole (1815–64), Alan Mathison Turing (1912–54) and John von Neumann (1903–57). Scientists build on what has gone before, which demands familiarity with the history of their own disciplines.

It is important to distinguish between 'information' and 'knowledge'. Information may consist of data, records of observations made and experimental results, but its value resides in the interpretation, which accumulates as knowledge. Knowledge, then, builds theories, which are descriptions of the universe and explanations of the phenomena within it and, according to Karl Popper, it is the development of theories that defines scientific progress.

As they develop, the theories tell us more and more, while at the same time excluding more and more, and the exclusions increase the opportunities for falsification. This strengthens the theories and leads Popper to refine his definition of progress: 'This consideration led to a theory in which scientific progress turned out not to consist in the accumulation of observations but in the overthrow of less good theories and their replacement by better ones, in particular by theories of greater content. Thus there was competition between theories – a kind of Darwinian struggle for survival.'[3]

An overturned theory is not necessarily erroneous. A crew of astronauts returning from the Moon were once asked, by Mission Control, who was navigating. 'Newton,' they replied. No one was directing their craft; it

232

automatically followed the orbit described by the Newtonian laws of gravity. Newtonian concepts were not disproved by those of Einstein, but augmented. The old ideas were not wrong, merely incomplete, and astronauts can continue to trust them. In the Popperian sense, the theory had developed through the acquisition of more content: it now tells us more than it did previously. Some theories are completely overthrown and abandoned, of course, because the more comprehensive explanation that replaces them also falsifies their central tenets. But, as was pointed out earlier, no opprobrium attaches to the author of an honest mistake (provided he or she does not make too many!).

People sometimes talk loosely of 'scientific disasters'. Obviously, there can be no such thing, unless that is how we choose to describe a theory that comes to be rejected. Many industrial processes exploit scientific principles. Should those processes suffer catastrophic failure, the result may be an industrial disaster, but it is meaningless to call it a scientific one. If the responsibility for the disaster is attributed to individuals, they will be managers, operators of machines, maintenance workers or perhaps the designers of the machinery, but not the scientists who discovered the principles. It would be a truly scientific disaster if, and only if, a factory was built and then failed to work because the scientific principles it sought to exploit were incorrect. Such occurrences happen rarely, if at all, and they cause no physical injury.

'Knowing the mind of God'

If scientific progress consists in the formulation of increasingly comprehensive and robust descriptions and explanations of phenomena, it is not too difficult to predict how scientists can run into trouble. Unless their enterprise fails and progress is checked, sooner or later they will be drawn into theorizing about topics on which people hold and express strong opinions. The search is for all-embracing theories, for complete explanations of all phenomena, and it has a long history. All the major religions, and probably many lesser ones, provide such explanations, and it is tempting to see them as in competition with each other. After all, they are often entirely different from and contradict one another. Some even contradict themselves, so must be false. If scientists can devise an alternative, why should they spare its rivals the rigorous attempts at falsification to which scientific explanations are subjected?

But scientists are human, and some of them can be expected to enjoy controversy and even to take delight in shocking more sensitive souls. No doubt, Stephen Hawking derives many hours of quiet amusement from the sport of Hawking-bashing he has done so much to encourage, especially when his enraged opponents are often so wide of the mark. Hawking is most commonly taken to task for his suggestion that one day we may 'know the mind of God', and in a review of his most recent book, Appleyard describes as nonsense Hawking's description of physics as the most fundamental of disciplines.[4] However, this is no extravagant claim. It is simply a statement of fact and has nothing to do with personal prestige. Physicists do study the composition of matter at its most basic, or fundamental level. 'Knowing the mind of God' is more intriguing, but this, too, can be justified in terms of creation myths. These usually involve one or more gods and some explain the motivation that led them to embark on their creation. Cosmologists seek a scientific theory to explain the origin and eventual fate of the universe. The parallel is obvious even if the language used to express it seems contentious to those unaccustomed to having their own theories challenged.

Perhaps Hawking is being playful, but his remark is hardly trivial. Our traditional creation myths are obsolete and suitable only as stories for entertainment. On no account are they to be believed, because in most cases they are contradicted by evidence we can gather for ourselves. We would benefit greatly from their replacement by a plausible, robust theory that can be defended rationally, supported by observation and explained to people of every cultural tradition. Interestingly, Stephen Wolfram, a computer scientist, believes the 'theory of everything' may take the form not of a series of equations, but of a computer program.[5] If he is right, the program itself may not be easy to follow, but anyone with adequate computing power will be able to run it, presumably to provide graphic demonstrations of the way each aspect of the universe behaves.

Such a program might acquire some kind of scriptural authority among those who would view its explanations on their screens in the same spirit of reverence as others might read sacred texts. Probably, this danger cannot be averted, but it would be sad if the tentative character of all scientific descriptions were to be transformed into dogma, for the program would be subject to regular updating as new discoveries modified its detail. Given its importance, once someone had purchased a copy, perhaps updated versions might be supplied free and automatically. It might even be possible to destroy earlier versions by remote control. This is not being frivolous. The questing spirit that is so essential a source of our creativity is seriously inhibited by too much certainty. We need the search to continue.

Physics may be the most fundamental of disciplines, but it is not the only one. At a higher hierarchical level, biologists are also working towards a comprehensive theory of living organisms and their origin. Demonstrating it is likely to be even more controversial, since it will involve producing a living organism from inert materials and, once again, researchers may expect to hear outraged accusations of 'god-like' behaviour. Nor is it beyond the bounds of possibility that we will be introduced to machines that are truly intelligent. At the very least, they will pass the 'Turing test', in which a human communicates with another human and a machine, both of which are out of sight. The experimenter asks questions, makes remarks and generally conducts one side of an ordinary conversation. Some of the replies are from the other human, some from the machine. The machine passes the test if the experimenter cannot distinguish between them; the machine is then held to be indistinguishable from a human. Ironically, to pass the test the machine might need to suppress some answers, for example to questions involving arithmetical calculations, which it would find much easier than the human, and it would have to lie if asked whether it were human (as would the human if asked whether he or she was a machine).[6]

The quest continues. It is a quest for knowledge, for comprehension, and the fear it generates is understandable but misplaced. It is understandable because it subverts previous descriptions of the universe and so disorients us. Mary Midgley explains it as being lost: 'what is disastrous is not just ignorance or even error as such, but ignorance and error about the whole and where we stand in it – failure to understand the world sufficiently to grasp our own position in relation to what matters in it.'[7] We feel lost because the inadequacy of our old ideas has been revealed to us. This is also why the fear is misplaced. Humpty Dumpty cannot be reassembled, no matter how many horses and men the king sends to his rescue. Once we recognize as absurdly impossible the idea that the Earth rides on the back of a giant turtle, nothing can restore its lost credibility. Scientists offer us progress, and only if we embrace it can we hope to be led to the new, comprehensive and credible 'map' we need if we are to locate ourselves.

For some, the pursuit of knowledge is deeply suspect, mainly because of a remark made more than three centuries ago by Francis Bacon (1561–1626). In his *Religious Meditations*, he wrote: '*Nam et ipsa scientia potestas est* [knowledge itself is power].' This strikes a modern reader as an expression of machismo, of an urge to dominate and control, power to overpower as it were. That may indeed have been what Bacon had in mind, for he was concerned with the exploitation of resources for the benefit of humans. But attitudes change as the centuries pass, and today we can attach a quite different meaning to his statement. We still exploit resources, as do all species, but increasing knowledge allows us to identify alternative resources and alternative and less damaging ways of exploiting them. We can think of knowledge as empowering us not to dominate but to comprehend, not to control but to delight in observing what we comprehend, not to overpower but to manage subtly and with sensitivity. Manage we must, because together with power our knowledge has imposed on us responsibility for our actions. Scientific progress no longer offers the conquest of nature. It has proceeded far beyond such an outworn idea. The conquest we are offered now is of our own fear and sense of disorientation. It is an offer that it would be most foolish of us to refuse.

24

Science and Ethics

For many years, scientists involved in the breeding of farm livestock have employed a technique in which an embryo is physically divided into two or more identical parts. The division is made when the conceptus consists of two or four cells (blastomeres) within a protective membrane (the zona pellucida). At this stage the blastomeres are identical, differentiation to form particular tissues having not yet begun, so if they are teased apart to form two groups of one or two cells each, then re-enclosed within a donor zona pellucida, the blastomeres will continue to divide and both conceptuses will develop normally.[1]

In October 1993, scientists at the George Washington University Medical Center in Washington DC reported their use of the same technique to divide 17 human embryos as part of their research into fertility. The purpose was to increase the chance of a successful pregnancy by increasing the number of embryos available for implantation. The embryos they used were genetically abnormal and could never have survived; after separation they continued to divide, some of them five times, by which time they consisted of 32 cells. Then they all died.[2] At once, it was felt necessary to broadcast ethical disclaimers, making the point that scientists had no intention of cloning humans. Jeremy Rifkind accused the scientists of doing work of no redeeming value but simply to see whether it would work (an unlikely motive since the technique is well established). His Foundation on Economic Trends immediately began lobbying for the cloning of humans to be made illegal throughout the country.

Theoretically, cloning is feasible and its feasibility does raise ethical concerns. Practically, however, it is improbable. It is possible to produce a large number of genetically identical embryos by repeated divisions, but if those embryos are to become human beings they must be nurtured within a human body. One mother might carry several, but the number is limited, and if more clones are to be produced other women must volunteer to carry them. We are a long way from being able to raise foetuses outside the human body and since this would be ethically unacceptable it is doubtful whether anyone is contemplating it. Thus, the cloning of humans encounters a barrier over the recruitment of mothers, allowing the ethical issue to be resolved at a very early stage. That is the issue of surrogacy, and it has already been addressed. In any case, unless somewhere there is a man who wishes to father a large number of identical offspring, cloning is pointless. As we are so often told, there is no shortage of human beings!

Certain scientific research and its application do indeed raise ethical issues, but it is important to distinguish between genuine concerns and those that are illusory because the scientific process has been misunderstood. There is, for example, a routine test for the viability of human sperm. The zona pellucida (outer membrane) is removed from a hamster ovum and exposed to the sperm; if it is penetrated by spermatozoa there is a good chance the male is capable of fathering a child. When news of this procedure was mentioned in the British House of Commons some members fiercely denounced it, talking of human–hamster hybrids and other monsters. Their anger subsided when the technique was explained to them, but it demonstrated the ease with which ethical outrage can stem from misunderstanding.[3]

Grave concerns were also expressed over the use of human embryos in research into human fertility. Most of these centred either on the point at which an embryo becomes a person or on the point at which a human life commences. Unfortunately, there is no simple answer to these questions. Life does not commence at all in a biological sense, since all living cells are in the condition of being alive and if the cells are from a human, the life characterizing them is also human. Nor is it any easier to define the point at which 'a person' is formed, except as the quite early stage at which a foetus is capable of surviving independently. Further confusion arose from the mistaken supposition by some people, that the research involved well-developed foetuses, rather than embryos at a very early stage in their development. The matter was resolved by licensing the use of embryos up to fourteen days after fertilization, which is the end of the implantation period wholly devoted to the laying down of tissues to nourish and protect the foetus, which has not yet begun to form.

Animal experiments

It is experimentation with non-human animals that arouses the strongest feelings. The feelings are justified, but often the issues are more complex than they seem. Opposition of this kind is not a recent phenomenon. Mary Midgley quotes Voltaire on the subject: 'You discover in it [the experimental animal] all the same organs of feeling that are in yourself. Answer me, mechanist, has nature arranged all the means of feeling in this animal so that it may not feel?'[4]

The first organized opposition to vivisection seems to have begun around 1870, mainly in Britain. However, the first society may have been founded by the wife and daughters of the French physiologist Claude Bernard (1813–78). They came home one day to discover that he had dissected the family dog.[5] He is said to have 'described the "science of life" as a superb salon resplendent with light which could be attained only by way of a long and ghastly kitchen.'[6] At least, he admitted that the kitchen was 'ghastly'. If the cruelty were wanton the matter would be simple, but Bernard made many very important discoveries about digestion, the function of the liver, tissue respiration, the nervous system and therapeutic uses of paralysing poisons. It was he who first proposed that living organisms have to maintain a constant internal environment (homoeostasis) because cells can function only within certain limits of temperature and osmotic pressure.

Bernard was far from insensitive to the suffering of animals. 'Our imagination cannot conceive of anything more wretched than that sensitive beings, who feel pleasure and pain, should be unable to seek one and avoid the other,' he wrote of his work with curare, which paralyses without desensitizing.[7] But he strongly defended the right of humans, who make use of animals for work and food, to use them in the acquisition of biological understanding and in the testing of therapies. Galen of Pergamum (130–200) is believed to have been the first physiologist to dissect live animals, especially pigs and monkeys, but in his day vivisection was long established; earlier, it was condemned criminals who were vivisected.

Extreme positions can be adopted on either side of the debate, but absolutism contributes little to finding a resolution. The choice is not between unrestricted experimentation and its total abandonment, but at what point we locate the unavoidable compromise. This turns on the moral worth we ascribe to non-humans, essentially a philosophical issue that arises partly from discoveries resulting from the very research under consideration. These discoveries have removed what was once a sharp distinction between the relative capacities of humans and non-humans to experience pain, frustration and deprivation and to possess intentions. Animals can suffer in ways recognizable to humans and are not mere automata. The significance of this was pointed out by Jeremy Bentham, the Utilitarian philosopher, in his *Introduction to the Principles of Morals and Legislation*, published in 1789. 'The question is not, Can they *reason?*' he wrote, 'nor Can they *talk?* but Can they *suffer?*'[8]

For those who would express their compassion only for species that seem intelligent, Marian Stamp Dawkins has argued persuasively that intelligence and capacity for suffering are inextricably linked and no animal could relate to its environment well enough to survive unless it possessed both in some degree.[9] It is also clear, as Phyllis Grodsky wrote some years ago, 'that the class of moral agents is far smaller than the class of moral objects, and that arguments about the moral equality of animals and humans concern the latter status and not the former.'[10] Our clear duty towards non-humans conflicts with their indispensability in some areas of research, producing a grey area within which we must seek our compromise. It is a quest that research scientists are actively engaged in. Few biologists would defend the use of animals in the testing of cosmetics, for example, but most would agree that animals continue to play an essential role in biomedical research and must probably continue to be used in toxicity testing.

Considerable progress has been made in these areas. In many countries, including Britain, it is illegal to perform unlicensed experiments using vertebrates. Licences are awarded to individuals, not institutions, for specified procedures that cannot be performed without the use of animals and are judged likely to contribute usefully to human or animal welfare. Licensees are directly accountable for their work and conditions under which it must be performed, and their animals housed, fed and treated, are strictly specified. Their premises may be inspected at any time, without warning, on behalf of the licensing authority. Minimum standards are also laid down within the European Union and by the Council of Europe, and on a worldwide basis by the Council of International Organizations of Medical Sciences.

If we can resolve the problem of our treatment of non-humans, we may be able to apply that solution to related ethical problems that have not yet arisen, but may do so in the future. We should, for example, be able to clarify our response to a hypothetical encounter with extraterrestrial beings. If the species is self-evidently intelligent, we may think there is no problem. We could behave towards it as we would behave towards other humans. This supposes that we treat all humans in the same way, of course, which is untrue, and we should remember that an extraterrestrial would almost certainly be a great deal less like you and me than a member of any human minority. In any case, it may not be easy to recognize intelligence in a being so different from ourselves. If the species seems unintelligent, plant-like perhaps, we might think it satisfactory to regard it as we would an equivalent terrestrial species. This could mean exterminating or conserving it depending on which seemed more appropriate.

Then again, should we one day be convinced of the sentient intelligence of a machine, we might feel a duty to ascribe some moral worth to it. We might think it impermissible simply to smash it to pieces, for example, or to torment it. Like Frankenstein, we might have to respond to its demand for a companion to bring it solace, and we might hope our response would be more compassionate than his. These are remote possibilities, of course, but not absurd ones. We are, after all, actively searching for evidence of extraterrestrial life and seeking to develop intelligent machines. It would be prudent to forearm ourselves for the implications of possible success and in this the debate over animal experimentation may well prove helpful.

How accountable?

Broader concerns centre on the issue of how far scientists can be held accountable for the application of their discoveries, it being suggested that certain lines of research should not be pursued because of the likely consequences of those applications. For example, chapter 3 mentioned work on new drugs that will alter not only mood, but our perceptions of ourselves. Before such substances are prescribed we should consider whether we deem it wise to interfere at this level and, in light of our deliberations, offer ethical advice to those who may be charged with the task.

As far as it affects research, at one level the matter is simple. Scientists are members of society and informed by the same moral considerations and attitudes as non-scientists. They will not undertake work they consider immoral and any who would be prepared to do so would be censored by the scientific community. Projects that are clearly immoral will not be funded and if, by any chance, they do attract private funding the results will not be accepted for publication. This was demonstrated some years ago by fierce arguments among scientists themselves over the permissibility of using data gathered by Nazi scientists that had been obtained by performing cruel experiments on humans. The data were valuable, and obviously unique, but because they had been obtained by unethical means many people felt these tainted any use of them.

In this case the controversy concerned only scientists and was confined to them. Where non-scientists become involved the issues are usually more complex. With hindsight it is easy to criticize decisions because of consequences that occurred long after the decisions were made. People have asked, for example, whether physicists should have agreed to develop nuclear weapons. To answer the question it is necessary to understand that the scientists had good reason to believe such weapons were being developed in Germany, and might be used as soon as they were ready. It was a deadly serious race at a time when their country was at war. Once the bombs had been made, it hardly seems fair to blame the physicists for having lost control of them. Similarly, present-day scientists whose research leads directly or indirectly to the development of weapons work in the belief that those weapons are necessary and desirable for the good of their own societies. Not everyone will agree with them, but neither will everyone disagree and, quite properly, policies on weapons development are decided by politicians, not scientists. It would be profoundly undemocratic for scientists to hold society to ransom by uniting to prohibit all work they considered undesirable, but which most people and their governments wanted.

From time to time, scientists are also blamed for discoveries that are applied in ways that lead to environmental damage. Ironically, it is usually scientists who identify that damage and draw attention to it, and they who devise ways to end the abuse. When news of the damage is made public, the commentators sometimes forget to mention any benefits that may have been derived from the damaging activity. Harm caused by pesticides, for example, should be weighed against the huge contribution they make to human health and food production. In the case of the Chernobyl accident, western nuclear scientists pointed out many years ago that the Soviet RMBK design was unsafe because it uses a graphite moderator through which pipes are routed carrying cooling water; at high temperatures and pressures graphite and water can react explosively, as they did. No western reactor was built to this design.

The increasing emphasis that politicians put on the economic benefits resulting from scientific research favours applied rather than fundamental research and makes it harder for scientists to establish their credibility. Presented with a statement from a scientist working for a commercial enterprise, people are inclined to think that it might be coloured by commercial considerations. This scepticism extends to others. Government scientists are suspected of defending political expedience, and academic scientists of protecting the commercial or political source of their funds.

The effect is insidious and can be countered only by building an atmosphere of trust between scientists and non-scientists. From the generally anti-scientific position our society has reached, this will not be easy, but it is vital if we are to deal rationally with the ethical problems we perceive. Scientists hold moral views, like everyone else, and are actively engaged in raising and discussing ethical issues of public concern, issues that can be resolved only through a full understanding of their scientific dimensions.

The way to that trust lies in increasing the amount of fundamental research and in working strenuously to explain to non-scientists what scientists do and why they do it. A common treatment for people suffering phobia is to expose them, step by small step, to whatever may be the object of their fear so they can learn from experience that the danger is unreal. A similar treatment might cure us of our phobia about the scientific enterprise.

25

Reconciling the Two Cultures

'Curiosity-driven science is somehow frivolous, and a luxury we can no longer afford.' Roy Schwitters, director of the project to build the Superconducting Super Collider (SSC), was understandably bitter at the cancellation of the project by the US Congress in October 1993.[1] Much of it had already been built and about half of the two thousand workers on the project had moved to the Texas site because of it. Admittedly, the SSC was expensive. The $11 billion it was estimated to cost is a great deal of money, even in the United States.

Although the SSC has been killed, the work planned for it will continue elsewhere and by slightly different means. CERN, the European research centre for particle physics, is seeking answers to the questions the SSC would have asked. These concern the ultimate composition of matter and, from that, the composition, structure and evolution of the universe. It is expensive because the particles physicists need to study can be produced only by colliding with other particles at extremely high energy. The SSC had a 24-km (15-mile) circular underground tunnel containing pipes through which particles would be accelerated by superconducting magnets. The tunnel will now be sealed.

Desire to reduce the federal deficit was only part of the reason for cancellation, however. Like most governments, the US government increasingly values its scientists in financial rather than intellectual terms, for the wealth they can create and the employment they can generate, showing scant interest in the 'mere' advance in our understanding of the world. Schwitters was right: curiosity about ourselves and the universe is frivolous.

The curse of short-termism

High-energy physics can answer questions, but almost certainly that is all it can do. It can contribute to our understanding of the universe we inhabit and, perhaps, our own origin. It can help us locate ourselves. These aims are now held to be of little importance. No one will make a fortune from them, no factories will be built, stock markets will take no notice. We live in an anti-intellectual and, more specifically, anti-scientific culture. Questions scientists ask can be answered more cheaply by pure reason or pure speculation. The answers may be wrong, but they will satisfy the idly inquisitive and in the absence of rigorous investigation no one can challenge them. It would be hard to devise a more short-sighted policy or one less likely to make any serious contribution to human happiness or welfare. Even in its own terms it is deeply flawed, and those terms are in any case utterly trivial.

Many, perhaps most, of the technological advances that have proved economically important in modern times began as curiosity-driven research. Computing is based on mathematical principles, abstract as abstract can be. In 1847, the professor of mathematics at Queen's College Cork, George Boole (1815–64), managed to bridge mathematics and formal logic by means of a new form of algebra, known today as 'Boolean algebra', which allows symbols to be attached to logical operations and then manipulated according to fixed rules. Boolean algebra underlies all programming of digital computers, but that is not why Boole invented it.

Genetic engineering is built on the discovery of the structure of nucleic acids made, in 1953, by Francis Crick (b. 1916), a physicist specializing in X-ray crystallography, and James Watson (b. 1928), a zoologist. They were motivated by curiosity, for no one could have predicted their work might have industrial applications. The whole of nuclear physics is derived from investigations of the structure of matter. So far from foreseeing its large-scale industrial (or military) applications, Lord Rutherford (1871–1937), generally regarded as the founder of the discipline, described the suggestion that it might prove useful as 'all moonshine'.

Financiers find this highly unsatisfactory. It is all very uncertain. The curiosity of scientists may sometimes prove economically profitable, but more often it does not. All that their self-indulgence generates is information, which is of little value to anyone, with the possible exception of publishers and printers. In any case, the lead times are much too long to guarantee an acceptable return on capital invested. It makes more sense to hire some specialists, set them to work inventing potentially profitable gadgets and fire them if they fail to produce results quickly enough.

In their pursuit of short-term profit, however, the financiers may deny us, and themselves, the opportunity not simply to produce novelties but to devise entirely new industries. Those who equate satisfying curiosity with self-

245

indulgence close their eyes to possibilities of advances so radical we cannot imagine the benefits they may bring until they are almost upon us. What is more, we need those benefits urgently if we are to construct a tolerable future for the majority of humans. Jesse Ausubel, director of the Program for the Human Environment at Rockefeller University, has summarized that need. 'If many people are to eat, dwell and move better than they do today, meeting the needs of 2020 will entail at least a doubling of agricultural production, energy services, industrial production – and a billion new jobs. If such growth is achievable, it might not be sustainable, at least with current technologies.'[2] We must develop new technologies and to do so we will require scientific research that is basic, or curiosity-driven, as well as applied.

Of course, our physical needs must be met, but humans also have psychic needs that cannot be overlooked. Scientists are often accused of pursuing economic gain, a criticism more properly levelled at their masters. The truth is that no group in society has contributed so much to the search for answers to the deeper questions that trouble or intrigue us all. It is in the dismissal of this role for scientific research that the emphasis on wealth-creation is so fundamentally misguided.

All humans are curious, we all need to locate ourselves, and it is from the scientific search that reliable answers are most likely to emerge. So many answers have been supplied already that we can no longer be satisfied with speculation. There is a real chance that scientists may be able to discover the answers to some of the most searching questions: how life originated and how our own species evolved; how the universe began, whether or not it will end and, if so, when; whether we are the only intelligent beings, and our planet the only one, out of the countless billions that must be orbiting other stars, to support life of any kind. The story is but partly told and we cannot leave it there, for the rest to be fabricated by fools and charlatans. It is too important for that, and to reduce scientific research to no more than widget design is to trivialize both it and ourselves.

The two cultures

On the evening of 7 May 1959, Sir Charles (C.P.) Snow delivered the annual Rede Lecture in the Senate House of the University of Cambridge and fuelled the debate over these issues that has continued ever since. In his lecture, 'The Two Cultures and the Scientific Revolution', Snow exposed and explored the gulf that had emerged between scientists and literary non-scientists. 'The non-scientists have a rooted impression that the scientists are shallowly optimistic, unaware of man's condition,' he said. 'On the other hand, the scientists believe that the literary intellectuals are totally lacking in foresight, peculiarly unconcerned with their brother men, in a deep sense anti-intellectual, anxious to restrict both art and thought to the existential moment.'[3]

The lecture provoked the eccentric critic F.R. Leavis into launching what was considered to be a disgraceful denunciation of its theme and of Snow personally. But despite the controversy that this attack provoked, even requiring Snow to defend his use of the word 'culture' to describe scientific work and understanding, the lecture achieved little of practical value. In the decades that have elapsed since, the gulf has continued to widen. If it seemed necessary to bridge it then, now it is critical, for the triumph of the anti-scientific branch of our culture threatens to close our minds and foreclose all our most hopeful options for the future.

Describing invective against science as destructive, Snow went on to address the matter of scientific optimism, the 'accusation that has been made so often that it has become a platitude'. It arose, he said, from a confusion between personal and social attitudes. Personally, most scientists consider the condition of the individual human being to be tragic. We are alone, seeking escape from our solitude, and we die alone. There is no reason, however, why the tragedy of the individual should imply social tragedy. Socially, there are problems that can be solved, answers that can be found, and it is in this consideration that scientific optimism consists. A scientist friend once told me the reason he was drawn to scientific research was the opportunity it provided to find answers to his questions. That is the optimism of the scientist.

In uncertain times it is optimism that we need. Unless they are challenged vigorously, the pessimists who rejoice in their predictions of apocalypse may prove themselves right. Their prophecies may prove self-fulfilling because of the despair they have engendered. If the future, as presented to us, is unrelievedly dismal, the only sensible thing to do is ignore it and live for the present, in what Snow called 'the existential moment'. So we reduce our lives to a daily struggle for the profit that will pay for whichever route to oblivion we choose, meanwhile imploding psychologically into an intro-spection that leads us to suppose that reality resides only in subjectivity, a new solipsism. Truth being subjective and therefore relative, one description of the world is as good as another. We can choose a ready-made description, scientific or otherwise, or devise one for ourselves. It makes little difference.

Thus beliefs flourish that were once called superstitious and are held with a kind of dogmatic sentimentalism. The invented world harbours magicians and spirits, the planet becomes a benign mother whose delicate sensibilities must not be affronted, animals are either depicted as cuddly children or ignored, and only humans are intrinsically evil. We contemplate equably the prospect of our own, richly deserved, extinction. This is the pattern of anti-scientific pessimism and its familiarity marks the extent of its triumph.

Reconciliation between scientists and others can be achieved only through compromise and a willingness to learn. Snow caricatured two groups of people each holding the other in contempt, so the educational process might begin with a relaxation from entrenched positions. For their part, scientists might question the widely held assumption that the arts and humanities are an intellectual 'soft option'. Although it cannot be quantified, painting and acting demand great precision compared with which scientific statements can seem vague. These skills and disciplines have value. The arts enrich our lives and the humanities contribute much to our understanding of society and our well-being.

The greater movement must come from non-scientists, however. Scientists, as people, are exposed to the arts and are as likely as anyone else to be politically and socially informed. Only the most determined hermit can hide from books, paintings, music, television and newspapers. Most of us would not choose to do so and scientists experience and enjoy these things, simply by virtue of being functioning members of our human society. Their own work and ideas can remain hidden, however. Someone ignorant of them exhibits no evident handicap and feels no deprivation. All of us see books and paintings and hear music, so we may be drawn to discover more of the delights they can offer. We are not exposed to scientific ideas in the same way. These offer delights at least as great, but they are not so immediately accessible.

A first step might be for newspapers to allocate more space and broadcasting more time to scientific news. If there really are two cultures, one scientific and the other non-scientific, then surely equal attention should be given to both. To do otherwise is to do a disservice to the public. The coverage of scientific topics should occupy a similar amount of space and time as are given to coverage of literature and the visual and performing arts. Indeed, a case could be argued for 'positive discrimination' in favour of scientific items to redress many years of imbalance. Surveys have identified a strong public demand for more scientific information, so the reform would not amount to an unpopular imposition, and any science journalist or author will confirm that there is no shortage of suitable material. Countless exciting stories await the telling and, handled well, they would have high entertainment value.

The eventual aim would not be the presentation of one 'culture' as superior to another, but their unification through making them both equally familiar to us. Until non-scientists can converse as comfortably about, say, cosmology or molecular biology, as cosmologists and molecular biologists do about novels or music, we can neither claim to be an educated, whole society nor can we embark on the next, more difficult step. This will require the unified culture to engage the problems generated by what has now become a third culture, composed of anti-rationalists and anti-scientists. Confronted with scientifically founded information about an objective reality whose existence all members of the newly unified culture accept, neosolipsism, 'new ageism', and cosy sentimentalism can be exposed as the vapid nonsenses they are.

Hope through reconciliation

If we achieve this much, a most significant change will have taken place. It is not merely that more of us will know and understand more about what scientists are doing, valuable though that will be, but to some extent society at large will be influenced by what is loosely called the 'scientific method'. This is a troublesome term, for in truth there is no such thing as a single method employed by all scientists, but many methods devised to suit the particular phenomena that individual scientists have chosen to examine. Yet there is an attitude that all scientists share: an attitude of criticism. Scientists are by nature sceptics. Their response to assertions is to demand evidence, logically consistent arguments, suggestions for ways in which those assertions might be tested – by which they mean falsified. It is the attitude, or method, that has led to the accumulation of everything they have revealed to us, and the popular adoption of a similarly rigorous approach would have profound implications. It will not turn all of us into scientists, because scientists practise professions for which long, arduous training is necessary, but it will imbue all of us with an important element of the 'scientific attitude'. Freed from cant, we will be able to address seriously the real problems of the world. Some may defeat us, but most are soluble and none threatens the survival of humanity, far less that of the planet itself.

So we will advance, at times slowly perhaps and with temporary reversals, but always with hope based on the optimism with which scientists have infected us. As old problems are solved new ones will emerge, but we will be well enough informed to recognize that humans have always faced challenges and always will, and no longer will we allow ourselves to be seduced into finding scapegoats to blame and using them as an excuse for inaction.

As time goes on we may once again take to calling our advance 'progress', because we will see that it leads towards a future that is better than the past, in which people can be better fed, better housed, better cared for and better educated than their parents. The historical experience of those of us fortunate enough to live in one of the industrialized countries can be repeated elsewhere, as it has been already in much of Asia. Our realistic appraisal of the world we inhabit will allow reasoned optimism to displace irrational pessimism.

We may also feel more secure, as the unification of cultures allows us to recognize the ability of each to answer questions posed by the other, and the answers to be understood. People may look back in bewildered amusement at those who today think scientists arrogant for turning their attention to the 'big questions' that many non-scientists hold to be beyond their competence, preferring speculation to calculation. Our increased security will be physical, but also psychological. We will have gone some way towards finding our place in the universe, so will feel less disoriented.

As the future appears less forbidding, we will lose our fear of it as a healthily growing child loses its fear of the dark. With that loss will go the decadence that marks our society today. Advancing beyond doom, we will have resurrected tomorrow and confidently claimed it as our own.

Notes and References

I THE ONE-EYED MAN
1 Scientists in Fiction

1 Gansberg, Alan L. 1983. *Little Caesar: A biography of Edward G. Robinson*. New English Library, Sevenoaks, Kent. p. 83.
2 Ibid. p. 81.
3 Nicholls, Peter (ed.). 1982. *The Science in Science Fiction*. Michael Joseph, London.
4 Rose, Hilary and Rose, Steven. 1969. *Science and Society*. Penguin Books, Harmondsworth. Introduction.
5 Rosen, Fred S. 'Pernicious treatment', review of *Lorenzo's Oil*. *Nature*, **361**, 695, 25 February 1993.
6 Shelley, Mary. 1818. *Frankenstein or, The Modern Prometheus*. Letter 4.
7 Ibid. Chapter 3.
8 Ibid. Chapter 4.
9 Ibid. Chapter 5.
10 In her introduction to the Standard Novels edition of *Frankenstein*, 1831.
11 Ibid.
12 The experiment is said to have been performed by a man named Crosse. Chambers reported it in his *Vestiges of the Natural History of Creation* (first published in 1844) and the story is recounted in Oldroyd, D.R. 1980. *Darwinian Impacts: An introduction to the Darwinian Revolution*. Open University Press, Milton Keynes.
13 Crichton, Michael. 1991. *Jurassic Park*. Random Century, London.
14 See Wilkins, Adam S. 1993. 'Jurassic hype', letter to *Nature*, **364**, 568, 12 August 1993.
15 See Nicholls, op. cit.
16 Christy, Campbell. 1984. *Nuclear Facts*. Hamlyn, London. pp. 33, 52.
17 Kahn, Hermann, Brown, William and Martel, Leon. 1976. *The Next 200 Years*. Associated Business Programmes, London. p. 220.

2 Scientists, the Barbarians at the Gates

1 Appleyard, Bryan. 1992. *Understanding the Present*. Pan Books, London. Chapter 8.
2 Quoted by Andro Linklater in 'Together again', an article on a Science Museum exhibition of scientific instruments. *Telegraph Magazine*, 30 October 1993.

3 Appleyard, op. cit.
4 Midgley, Mary. 1992. *Science as Salvation: A Modern Myth and its Meaning*. Routledge, London. p. 167.
5 The four physical forces are the electromagnetic force, the strong nuclear force, the weak nuclear force and gravity.
6 Hawking, Stephen. 1988. *A Brief History of Time*. Bantam Press, London. p. 175.
7 Appleyard, op. cit. pp. 1–2, 179; Midgley, op. cit. pp. 8, 89.
8 Dyson, Freeman. 1979. *Disturbing the Universe*. Harper and Row, New York. Chapter 23, 'The Argument from Design'.
9 Ibid.
10 Horgan, John. 1993. 'Perpendicular to the Mainstream', a profile of Freeman Dyson. *Scientific American*, August 1993. p. 16.
11 Polkinghorne, Revd Dr John. 1993. 'Can a scientist pray?' *RSA Journal*, CXLI, 5441, July 1993.
12 Appleyard, op.cit. Chapter 2.
13 Allaby, Michael. 1989. *Guide to Gaia*. Macdonald Optima, London. Chapter 11.
14 Midgley, op.cit. p. 214.
15 Davy, Sir Humphry. From *Sons of Genius*, in Heath-Stubbs, John and Salman, Phillips (eds.). 1984. *Poems of Science*. Penguin Books, Harmondsworth.
16 Radford, Tim. *Guardian*, 2 September 1993.
17 Horgan, op.cit.
18 Hovis, R. Corby and Kragh, Helge. 1993. 'P.A.M. Dirac and the beauty of physics'. *Scientific American*, May 1993. p. 62.
19 Dirac, Margit W. 1993. Letter to *Scientific American*, August 1993. p. 6.
20 Pauli quoted in Heisenberg, Werner. 1971. *Physics and Beyond: Memories of a Life in Science*. George Allen and Unwin, London. Chapter 7.
21 See Warnock, Mary (ed.). 1962. *Utilitarianism*. Collins/Fontana, London. Introduction.
22 See Alcock, John. 1979. *Animal Behavior: An evolutionary approach*. Sinauer Associates, Sunderland, Massachusetts. p. 400 et seq.
23 The dilemma is this. You and an accomplice commit a crime, are arrested separately and are held apart so you cannot meet or communicate. A detective tells you there is insufficient evidence to convict unless one of you confesses, implicating the other. If neither of you confesses, you both go free. If you confess, but your colleague does not, you will receive a very light sentence, your colleague a very heavy one. Should you both confess, you will both receive intermediate sentences. Finally, you are told the same offer has been made to your colleague. How do you respond?
24 See Axelrod, R. 1984. *The Evolution of Cooperation*. Basic Books, New York.
25 Nowak, Martin A. and Sigmund, Karl. 1992. 'Tit for tat in heterogeneous populations'. *Nature*, **355**, 250, 16 January 1992.
26 Capra, Fritjof. 1975. *The Tao of Physics*. Wildwood House, London; also 1982. *The Turning Point*. Wildwood House, London.
27 Horgan, John. 1993. 'Quantum Philosophy'. *Scientific American*, July 1992. p. 77.
28 Chiao, Raymond Y., Kwiat, Paul G., and Steinberg, Aephraim M. 'Faster than Light?'. *Scientific American*, August 1993, p. 38.
29 For a lucid explanation see Hey, Tony and Walters, Patrick. 1987. *The Quantum Universe*. Cambridge University University Press, Cambridge.
30 Zohar, Danah and Marshall, Ian. 1994. *The Quantum Society*. Bloomsbury, London.
31 See, for example, Davies, Paul. 1983. *God and the New Physics*. Dent, London.

32 In a letter to *The Guardian*, 24 March 1992.
33 Cupitt, Don. 'The pure bliss of life in an eternal flux'. *Guardian*, 4 September 1993.
34 Goldsmith, Edward. 1992. *The Way*. Rider, London.
35 Goldsmith, Edward. 1978. *The Stable Society*. The Wadebridge Press, Wadebridge, Cornwall, UK. p. 32.
36 Ibid. p. 34.

3 The Quest for Certainty

1 Appleyard, Bryan. 1992. *Understanding the Present*. Pan Books, London. p. 204.
2 See Gleick, James. 1988. *Chaos: Making a New Science*. Heinemann, London. p. 93.
3 Koyré, Alexandre. 1954. Introduction to Anscombe, Elizabeth and Geach, Peter Thomas (eds.). 1970. *Descartes Philosophical Writings*. Nelson's University Paperback, London. p. xix.
4 Descartes, René. 1637. *Discourse on the Method*, Book IV, in Anscombe, Elizabeth and Geach, Peter Thomas (eds.). 1970. *Descartes Philosophical Writings*. Nelson's University Paperback, London.
5 Humphrey, Nicholas. 1987. 'The inner eye of consciousness', in Blakemore, Colin and Greenfield, Susan (eds.). *Mindwaves*. Basil Blackwell, Oxford.
6 Restak, Richard. 1993. 'Brain by Design'. *The Sciences*, September/October 1993. New York Academy of Sciences.
7 Penrose, Roger. 1987. 'Minds, Machines and Mathematics', in Blakemore, Colin and Greenfield, Susan (eds.). *Mindwaves*. Basil Blackwell, Oxford.
8 Zohar, Danah and Marshall, Ian. 1993. *The Quantum Society*. Bloomsbury, London.
9 Crick, Francis and Koch, Christof. 1992. 'The Problem of Consciousness'. *Scientific American*, September 1992. p. 111.
10 Ryle, Gilbert. 1949. *The Concept of Mind*. Hutchinson, London.
11 See Ayer, A.J. 1956. *The Problem of Knowledge*. Penguin Books, Harmondsworth. Chapter 5.
12 Report in *The Guardian*, 26 July 1993.
13 Hamer, Dean, et al. 1993. *Science*, **262**, 321.
14 King, Mary-Claire. 1993. 'Sexual orientation and the X'. *Nature*, **364**, 288.
15 Dawkins, Richard. 1993. 'Don't panic; take comfort, it's not all in the genes'. *Daily Telegraph*, 17 July 1993.
16 Slakey, Francis. 1993. 'When the lights of reason go out'. *New Scientist*, 11 September 1993. p. 49.

II THE FLIGHT FROM REASON
4 The Denial of Progress

1 Plato, *The Republic*, Book VII. Trans. John Llewellyn Davies and David James Vaughan. 1919. Macmillan, London.
2 Ibid. Book VIII.
3 Aristotle, *Politics*, Book V, 'Pathology of the State'. Trans. John Warrington. 1959. Heron Books, London.

4 Three cycles are involved: the eccentricity of the Earth's orbit with a period of about 100,000 years; the tilt of the Earth's axis, with a period of 40,000 years; and precessional change with a period of 20,000 years. When these cycles coincide to bring radiation maxima and minima, climatic change occurs. Milankovich identified nine glacial episodes linked to nine radiation minima.

5 Gleick, James. 1987. *Chaos: Making a New Science*. Heinemann, London. p. 195.

6 Fukuyama, Francis. 1992. *The End of History and the Last Man*. Hamish Hamilton, London. Chapter 5.

7 Mumford, Lewis. 1961. *The City in History*. Penguin Books, Harmondsworth. p. 206.

8 Ehrlich, Paul R. and Ehrlich, Anne H. 1970. *Population, Resources, Environment*. W.H. Freeman, San Francisco. p. 354.

9 White, Lynn Jr. 'The Historical Roots of our Ecologic Crisis', published originally in *Science*, reprinted in *The Environmental Handbook*, De Bell, Garrett (ed.). 1970. Ballantine and Friends of the Earth, New York, and quoted in Allaby, Michael (ed.). 1989. *Thinking Green: An Anthology of Essential Ecological Writing*. Barrie and Jenkins, London.

5 Decadence and Fear of the Future

1 According to the *Shorter Oxford English Dictionary*. 1964. Clarendon Press, Oxford.

2 Hampson, Norman. 1968. *The Enlightenment*. Penguin Books, Harmondsworth.

3 Carlyle, Thomas. 1831. 'Diderot', in *Foreign Quarterly Review*, **22**, 1. Paris.

4 Fast, Howard (ed.). 1948. *The Selected Works of Tom Paine*. The Bodley Head, London. p. 98.

5 Paine, Thomas. 1794. *The Age of Reason: Being an Investigation of True and Fabulous Theology*, in Fast, Howard (ed.). 1948. *The Selected Works of Tom Paine*. The Bodley Head. London.

6 Campbell, Joseph. 1972. 'The Confrontation of East and West in Religion' in *Myths to Live By*. Viking Press, New York.

7 See Russell, Bertrand. 1961. *History of Western Philosophy*. 2nd ed. George Allen and Unwin, London. p. 571.

8 Hampson, op. cit.

9 Carson, Rachel. 1963. *Silent Spring*. Hamish Hamilton, London.

10 Schweitzer, Albert. 1954. *My Life and Thought*. 2nd ed. Guild Books and George Allen and Unwin, London. Translated from *Aus Meinem Leben und Denken*, published in Leipzig in 1931.

11 Ibid. Quoted in the Postscript by Everett Skillings.

12 See Mellanby, Kenneth. 1992. *The DDT Story*. British Crop Protection Council, Bracknell, Berkshire.

13 Carson, op.cit. Chapter 3, 'Elixirs of Death'.

14 Whelan, E.M. 1985. *Toxic Terror*, Jameson Books, Ottawa, Illinois. Quoted by Mellanby, op.cit., p. 86.

15 Schweitzer, Albert. 1947. *The Hospital at Lambaréné During the War Years*. The Albert Schweitzer Fellowship of America.

16 Appleyard, Bryan. 1992. *Understanding the Present*. Pan Books, London. Chapter 5, 'From scientific horror to the green solution'.

17 Marshall, N.B. 1979. *Developments in Deep-Sea Biology*. Blandford Press, Poole, Dorset. p. 33.

18 Allaby, Michael. 1992. *Water: Its Global Nature*. Facts on File, New York. pp. 198–200.

6 The Age of Aquarius

1 Campbell, Joseph. 1970. 'Schizophrenia – The Inward Journey', in Campbell, J. 1972. *Myths to Live By*. Viking Press, New York.

2 See, for example, Claridge, Gordon. 1987. 'Schizophrenia and Human Individuality' in Blakemore, Colin and Greenfield, Susan (eds.). *Mindwaves*. Basil Blackwell, Oxford.

3 Appleyard, Bryan. 1992. *Understanding the Present*. Pan Books, London.

4 According to *Alternative England and Wales*, published in 1975 by Nicholas Saunders. The system was called 'The Headless Way'.

5 Slakey, Francis. 1993. 'When the lights of reason go out'. *New Scientist*, 11 September 1993, p. 49.

6 Kapitza, Sergei. 1991. 'Antiscience Trends in the USSR'. *Scientific American*, August 1991.

7 Button, John. 1990. *New Green Pages*. Macdonald Optima, London.

8 Ibid. p. 180.

9 Shipley, Joseph T. 1968. *Dictionary of Early English*. Littlefield, Adams and Co., Totowa, New Jersey.

10 Used in this sense, the word bears no relation to the same word as used by chemists. Since all material substances are composed of compounds that can be generically described as 'chemicals', the idea of a 'chemical-free' substance is, quite literally, meaningless. Similarly, a compound is what it is regardless of its source. All biological organisms contain nitrogen compounds, for example, and the compounds are identical whether produced by bacterial or industrial fixation of atmospheric nitrogen.

11 Flusfelder, David. 1993. 'Role models in decadent days'. *Daily Telegraph*, 8 May 1993.

12 There are two, in the first and second chapters of the Book of Genesis, and they differ. Difficulties also arise over the use of the 'morning and the evening' of the first three days, when the Sun and Moon, which define 'morning' and 'evening', were not created until the fourth day. (There are also two differing accounts of Noah and the Flood.)

13 British press reports by Derek Brown (*Guardian*) and Anton La Guardia (*Daily Telegraph*), 13 August 1993.

14 *Did Man Get Here by Evolution or by Creation?* 1967. Watch Tower Bible and Tract Society New York. p. 126.

15 Ibid. Chapter 11.

16 For examples of this see Groves, Colin. 1992. 'Drowning in Superstition: Flood Myths and their Significance'. *The Skeptic*, Spring 1992. Australian Skeptics, Melbourne.

17 Buderi, Robert. 1990. 'Approval for degrees turned down'. *Nature*. **343**, 501, 8 February 1990.

18 Anderson, Christopher. 1992. 'Creationist victory'. *Nature*. **355**, 757, 27 February 1992.

19 See Gould, Stephen Jay. 1983. *Hen's Teeth and Horse's Toes*. Penguin Books, Harmondsworth. Chapter 21.

20 See, for example, Thulborn, Tony. 1985. 'Rot sets in in Queensland'. *Nature*, **315**, 89, 9 May 1985, and subsequent correspondence in *Nature*, **316**, 184, 18 July 1985, **318**, 100, 14 November 1985, and **320**, 9, 6 March 1986, much of it concerning the dimensions of Noah's ark.

21 Olroyd, D.R. 1980. *Darwinian Impacts*. Open University Press, Milton Keynes. p. 250.

22 Ibid. p. 252.

23 For a simple account of this phenomenon see Ridley, Mark. 1985. *The Problems of*

Evolution. Oxford University Press, Oxford. For a briefer but more technical account see Cockburn, Andrew. 1991. *An Introduction to Evolutionary Ecology*. Blackwell Scientific Publications, Oxford.

24 Campbell, Joseph. 1961. 'The Separation of East and West', in Campbell, op. cit.

25 Ibid.

7 Monetarism and Idealism

1 See Appleyard, Bryan. 1992. *Understanding the Present*. Pan Books, London, and Midgley, Mary. 1992. *Science as Salvation*. Routledge, London.

2 Fukuyama, Francis. 1992, *The End of History and the Last Man*. Hamish Hamilton, London.

3 Ibid. Chapter 31.

4 Appleyard, op.cit. Glossary.

5 Galbraith, J.K. 1992. *The Culture of Contentment*. Sinclair–Stevenson, London.

6 See, for example, Titmuss, Richard M. 1959. 'The Irresponsible Society', printed as Fabian Tract 323 and included in Titmuss, Richard M. 1987. *The Philosophy of Welfare*. Allen and Unwin, London.

7 Coghlan, Andy. 1993. 'Nice research, shame about the profits'. *New Scientist*, 14 August 1993, p. 4.

8 'Does science leave room for the soul?' Editorial comment in *Nature*, **356**, 729, 30 April 1992.

9 Rose, Hilary and Rose, Steven. 1969. *Science and Society*. Penguin Books, Harmondsworth. p. 240.

III FANTASIES OF A JUVENILE APE
8 The Juvenile Ape

1 Goodall, Jane. 1986. *The Chimpanzees of Gombe: Patterns of Behavior*. The Belknap Press of Harvard Univ. Press. Cambridge, Mass. p. 590.

2 Ibid. p. 591.

3 Ibid. p. 6.

4 Oldroyd, D.R. 1980. *Darwinian Impacts: An introduction to the Darwinian Revolution*. Open University Press, Milton Keynes. p. 194.

5 Huxley, T.H. 1906. 'Man's Relations to Lower Animals' in *Man's Place in Nature and Other Essays*. J.M. Dent and Sons, London.

6 Groves, Colin P. 1991. *A Theory of Human and Primate Evolution*. Clarendon Press, Oxford. p. 156.

7 Lewin, Roger. 1989. *Human Evolution: An Illustrated Introduction*. 2nd ed. Blackwell Scientific Publications, Oxford.

8 For a detailed discussion of heterochrony and its implications, see Groves, op. cit. pp. 56–59 and 310–14.

9 Terrace, Herbert. 1987. 'Thoughts Without Words' in Blakemore, Colin and Greenfield, Susan (eds.). *Mindwaves*. Basil Blackwell, Oxford. p. 129.

10 Goodall, op. cit. p. 41.

9 Golden Age and Noble Savage

1 See Campbell, Joseph. 1966. 'The Emergence of Mankind' in Campbell, Joseph. 1972. *Myths to Live By*. Viking Press, New York.

2 See Galbraith, John Kenneth. 1992. *The Culture of Contentment*. Sinclair-Stevenson, London. p. 80.

3 Russell, Bertrand. 1961. *History of Western Philosophy*. 2nd ed. George Allen and Unwin, London. p. 268.

4 Ibid. p. 602.

5 Dubos, René. 1980. *The Wooing of Earth*. Athlone Press, London. p. 3.

6 Sale, Kirkpatrick. 1980. *Human Scale*. Coward, McCann and Geoghegan, New York. p. 17.

7 Goldsmith, Edward. 1974. 'How to avoid Flixboroughs'. *The Ecologist*, **4**, 6, 202.

8 Goldsmith, Edward. 1988. *The Great U-Turn: De-industrializing Society*. Green Books, Hartland, Devon. p. 7.

9 Ibid. p. 11.

10 Quoted in Mason, John Hope. 1979. *The Indispensable Rousseau*. Quartet Books, London. p. 49.

11 Quoted in White, David Gordon. 1991. *Myths of the Dog-Man*. University of Chicago Press, Chicago. pp. 49–50.

12 Hemming, John. 1978. *Red Gold: The Conquest of the Brazilian Indians*. Macmillan London Ltd. p. 13.

13 More, Sir Thomas. 1516. *Utopia: The First Book of the Communication of Raphael Hythloday*.

14 Headland, Thomas N. 1990. 'Paradise Revised'. *The Sciences*, September/October. New York Academy of Sciences. p. 45.

15 Sahlins, Marshall. 1974. 'The Original Affluent Society'. *The Ecologist*, **4**, 5, 181.

16 Molony, Carol. 1988. 'The Truth About the Tasaday'. *The Sciences*, September/October. New York Academy of Sciences. p. 12.

10 Mother Earth, Wilderness and El Dorado

1 Sale, Kirkpatrick. 1985. *Dwellers in the Land: The Bioregional Vision*. Sierra Club Books, San Francisco. pp. 13–14.

2 Myers, Norman (ed.). 1985. *The Gaia Atlas of Planet Management*. Pan Books Ltd, London.

3 According to *The Writer's Handbook 1993*. Turner, Barry (ed.). 1993. Macmillan, London.

4 Joseph, Lawrence E. 1990. *Gaia: The Growth of an Idea*. Arkana, Penguin Books, London. p. 65.

5 Lovelock, J.E. 1979. *Gaia: A new look at life on Earth*. Oxford University Press, Oxford.

6 Allaby, Michael and Lovelock, James. 1983. *The Great Extinction*. Secker and Warburg, London.

7 Allaby, Michael and Lovelock, James. 1984. *The Greening of Mars*. Andre Deutsch, London.

8 Charlson, Robert J., Lovelock, James E., Andreae, Meinrat O., and Warren, Stephen G. 1987. 'Oceanic phytoplankton, atmospheric sulphur, cloud albedo and climate'. *Nature*, **326**, 655, 16 April 1987.

9   Westbroek, Peter. 1992. *Life as a Geological Force: Dynamics of the Earth.* W.W. Norton and Co., New York. p. 206.
10 Lovelock, James. 1989. *The Ages of Gaia: A Biography of Our Living Earth.* Oxford University Press, Oxford. Chapter 3.
11 Russell, Peter. 1982. *The Awakening Earth: The Global Brain.* Ark Paperbacks, London.
12 Quoted by Dubos, René. 1980. *The Wooing of Earth.* The Athlone Press, London. p. 11.
13 Thomas, Keith. 1983. *Man and the Natural World.* Penguin Books, Harmondsworth. p. 254.
14 Young, Arthur. 1808. *General Report on Enclosures,* Appendix III. Republished 1971 by Augustus M. Kelley, New York.
15 Cobbett, William. 1830. *Rural Rides.* Republished 1967 by Penguin Books, Harmondsworth. p. 411.
16 Grove, Richard H. 1992. 'Origins of Western Environmentalism'. *Scientific American,* July 1992. p. 23.
17 Keay, R.W.J. 1990. 'Presidential Address 1990: A sense of perspective in the tropical rain forest', *Biologist,* **37**, 3, 73, June 1990.
18 Dubos, op. cit. pp. 111–12.
19 In Cumming, W.P., Hillier, S., Quinn, D.B. and Williams G. 1974. *The Exploration of North America 1630–1776.* Paul Elek, London. p. 50.
20 Ibid. pp. 164–5.
21 Hemming, John. 1978. *Red Gold: The Conquest of the Brazilian Indians.* Macmillan London, London, p. 185.
22 Ibid. p. 223.

IV FAUST AND HIS DEBT
11 Science, Technology, and Wealth Creation

1   See, for example, Rees, Gareth. 1993. 'The longbow's deadly secrets'. *New Scientist,* **138**, 1876, p. 24, 5 June 1993.
2   See Blair, Bruce C. and Kendall, Henry W. 1990. 'Accidental Nuclear War'. *Scientific American,* **263**, 6, 19 December 1990.
3   Turco, Richard P., Toon, Owen B., Ackerman, Thomas P. et al. 1984. 'The Climatic Effects of Nuclear War'. *Scientific American,* **251**, 2, 23 August 1984.
4   Walgate, Robert. 1986. 'Politics before scientific advice'. *Nature,* **322**, 762, 28 August 1986.

12 Reductionism, Holism and the Curious Nature of Truth

1   Roszak, Theodore. 1973. *Where the Wasteland Ends: Politics and Transcendence in Post industrial Society.* Faber and Faber, London. p. 264.
2   Zohar, Danah and Marshall, Ian. 1993. *The Quantum Society.* Bloomsbury, London. Chapter 1.

3 Roszak, op. cit. p. 258.

4 Ibid. p. 400.

5 Appleyard, Bryan. 1992. *Understanding the Present*. Pan Books, London. p. 125.

6 Tansley, A.G. 1923. *Practical Plant Ecology*. George Allen and Unwin, London. p. 15.

7 Appleyard, op. cit. p. 89.

8 See Wolpert, Lewis. 1992. *The Unnatural Nature of Science*. Faber and Faber, London. Chapter 3.

9 See Medawar, P.B. and Medawar, J.S. 1978. *The Life Science*. Paladin Books, London. pp. 164–5.

10 Ibid. p. 166.

11 Veggeberg, Scott. 1993. 'Escape from Biosphere 2'. *New Scientist*, **139**, 1892, 24, 25 September 1993.

12 Both poems are quoted in Heath-Stubbs, John and Salman, Phillips (eds.). 1984. *Poems of Science*. Penguin Books, Harmondsworth.

13 Hampson, Norman. 1968. *The Enlightenment*. Penguin Books, Harmondsworth. pp. 116–17.

14 In an essay called 'A theory of population deduced from the general law of animal fertility', published in 1852.

15 Oldroyd, D.R. 1980. *Darwinian Impacts: An introduction to the Darwinian Revolution*. Open University Press, Milton Keynes. p. 206.

16 Quoted in Galbraith, John Kenneth. 1992. *The Culture of Contentment*. Sinclair-Stevenson, London. p. 80.

17 Ryder, Richard. 1993. 'Violence and machismo'. *RSA Journal*, **CXLI**, 5443, 706. October 1993.

18 Galton, Francis. 1883. *Inquiries into Human Faculty and Its Development*. Everyman edition, J.M. Dent, London.

19 Quoted in Wolpert, op. cit.

20 Ibid.

21 Gould, Stephen Jay. 1981. *The Mismeasure of Man*. W.W. Norton, New York.

22 See, for example, Zohar, Danah and Marshall, Ian, op. cit.

23 Capra, Fritjof. 1975. *The Tao of Physics*. Wildwood House, London.

24 Quoted in Slakey, Francis. 1993. 'When the lights of reason go out'. *New Scientist*, **139**, 1890, 49, 11 September 1993.

25 Horgan, John. 1993. 'Profile: Paul Karl Feyerabend'. *Scientific American*, May 1993.

26 Theocharis, T. and Psimopoulos, M. 1987. 'Where science has gone wrong'. *Nature*, **329**, 595, 15 October 1987.

27 Similar views were also advanced in the fifth century BC by Gorgias, a sophist, who was alleged by Aristotle (a century later) to have proposed that nothing exists; if anything does exist it is unknowable; if it exists and is knowable, knowledge of it cannot be communicated to others. Aristotle strongly opposed this view.

28 Magee, Bryan. 1982. *Popper*. Fontana Paperbacks, London.

29 Theocharis and Psimopoulos, op. cit.

30 Hawking, Stephen. 1988. *A Brief History of Time*. Bantam Press, London. p. 124.

31 Ibid. p. 125.

32 Midgley, Mary. 1992. *Science as Salvation: A Modern Myth and its Meaning*. Routledge, London. p. 28.

33 Ibid. p. 9.

34 Twain, Mark. 'The Damned Human Race: I. Was the World Made for Man?' in DeVoto, Bernard (ed.). 1962. *Letters from the Earth*. Fawcett Publications, Greenwich, Conn.
35 Theocharis and Psimopoulos, op. cit.

13 Scientists as Citizens

1 Proteins consist of amino-acid molecules linked together by peptide bonds (bonds between a carbon–oxygen (CO) group and a nitrogen–hydrogen (NH) group).
2 See Franklin, Carl. 1993. 'Did life have a simple start?' *New Scientist*, 2 October 1993.
3 See Maynard Smith, John. 1979. 'Hypercycles and the origin of life', *Nature*, **280**, 445, 9 August 1979.
4 In *Analytical Chemistry*, June 1961, quoted in Hitching, Francis. 1982. *The Neck of the Giraffe or Where Darwin Went Wrong*. Pan Books, London. pp. 66–7.
5 Hitching, op. cit. p. 135.
6 Davenas, E., Beauvais, F., Amara, J. et al. 1988. 'Human basophil degranulation triggered by very dilute serum against IgE'. *Nature*, **333**, 816, 30 June 1988.
7 Maddox, John, Randi, James, and Stewart, Walter W. 1988. '"High-dilution" experiments a delusion'. *Nature*, **334**, 287, 28 July 1988.
8 'Cold fusion causes frenzy but lacks confirmation'. *Nature*, **338**, 447, 6 April 1989.
9 Deuterium has an atomic nucleus comprising one proton and one neutron. Since it is the positively-charged proton that determines the chemical characteristics of the atom, the neutron being neutral, deuterium behaves chemically just like hydrogen, with a nucleus comprising one proton and no neutron. For this reason, deuterium is an isotope of hydrogen, sometimes known as 'heavy hydrogen'; 'heavy water' is deuterium oxide, a compound chemically identical to ordinary water but in which deuterium atoms substitute for hydrogen atoms.
10 Gleason, Daniel F. and Wellington, Gerard M. 1993. 'Ultraviolet radiation and coral bleaching'. *Nature*, **365**, 836, 28 October 1993.
11 See Lovelock, James. 1989. *The Ages of Gaia*. Oxford University Press, Oxford. Chapter 4.

14 Pollution and the Fear of Numbers

1 Cole, LaMont C. 1970. 'Can the World Be Saved?' in Johnson, Cecil E. (ed.). *Eco-Crisis*. John Wiley & Sons, New York.
2 Klotz, John W. 1972. *Ecology Crisis*. Concordia Press, London. p. 27.
3 Matthews, William H., Kellogg, William W., and Robinson, G.D. (eds.). 1971. *Man's Impact on the Climate*. MIT Press, Cambridge, Mass. p. 11.
4 British Medical Association. 1991. *Hazardous Waste and Human Health*. Oxford University Press, Oxford. p. 88.
5 Pearce, Fred. 1991. *Green Warriors*. The Bodley Head, London. pp. 62–3.
6 US Department of Health, Education, and Welfare. 1969. *Report of the Secretary's Commission on Pesticides and Their Relationship to Environmental Health*. US Government Printing Office, Washington DC. p. 14.

7 Ames, Bruce N. 1992. 'Facts Versus Phantoms'. *Perspectives*, **2**, 3, Winter 1992/3. Wells Associates for Guinness PLC, Lausanne.
8 The phenomenon and its dangers are clearly described by Toby Young in 'Little brother is watching you'. *Guardian*, 14 August 1992.
9 Ames, op. cit.
10 Press release WHO/9, 11 May 1993. World Health Organization, Geneva.
11 See Thomas, Keith. 1983. *Man and the Natural World*. Penguin Books, Harmondsworth. pp. 244–5.
12 UNEP/WHO press release, 1 December 1992.

15 The Malthusian Doctrine, or Blaming the Victim

1 Ehrlich, Paul R. 1970. 'Too Many People', extract from *The Population Bomb*, in De Bell, Garrett (ed.). *The Environmental Handbook*. Ballantine Books, New York.
2 *Guardian*, 22 September 1993.
3 Data from *Britannica Book of the Year*, 1974 and 1993.
4 The process is described clearly in Piel, Gerard. 1992. *Only One World: Our Own to Make and to Keep*. W.H. Freeman, New York. p. 284 et seq.
5 Malthus, Thomas Robert. 1798. *An Essay on the Principle of Population as it Affects the Future Improvement of Society*. Penguin Classics edition. 1982. Chapter 1.
6 Ibid. Chapter 5.
7 Tranter, Neil. 1973. *Population since the Industrial Revolution: The case of England and Wales*. Croom Helm, London. p. 41.
8 Luthi, Theres. 1993. 'If I Were a Rich Man'. *The Sciences*, November/December 1993. p. 5.
9 Grylls, Rosalie Glynn. 1953. *William Godwin and his World*. Odhams Press, London.
10 For a full account see Inglis, Brian. 1971. *Poverty and the Industrial Revolution*. Hodder and Stoughton, London.
11 Quoted in Inglis, op. cit. p. 74.

16 Food, Famine and the Depletion of Resources

1 Quoted in Burgess, G.H.O. 1967. *The Curious World of Frank Buckland*. John Baker, London. pp. 87–8.
2 Burnett, John. 1966. *Plenty and Want: A social history of diet in England from 1815 to the present day*. Nelson, London.
3 Borgstrom, George. 1972. *The Hungry Planet*. 2nd edition. Collier Books, New York. Introduction.
4 Tudge, Colin. 1977. *The Famine Business*. Faber and Faber, London. p. 2.
5 Meadows, Donella H., Meadows, Dennis L., Randers, Jørgen, Behrens, William W. III. 1972. *The Limits to Growth*. Earth Island, London. p. 9.
6 Ibid. p. 23.
7 Cloud, Preston. 1970. 'Mined out!' *The Ecologist*. **1**, 2, 25 August 1970.

8 Ehrlich, Paul R. and Ehrlich, Anne H. 1972. *Population Resources Environment*, 2nd ed. W.H. Freeman, San Francisco. p. 72.
9 Goldsmith, Edward, Allen, Robert, Allaby, Michael, Davoll, John, and Lawrence, Sam. 1972. 'A Blueprint for Survival'. *The Ecologist*, **2**, 1.
10 Mesarovic, Mihaijlo and Pestel, Eduard. 1975. *Mankind at the Turning Point*. Hutchinson, London.
11 Independent Commission on International Development Issues under the Chairmanship of Willy Brandt. 1980. *North-South: A programme for survival*. Pan Books, London. Chapter 11.
12 The World Commission on Environment and Development. 1987. *Our Common Future*. Oxford University Press. pp. 45–6.
13 Kahn, Herman, Brown, William, and Martel, Leon. 1977. *The Next 200 Years*. Associated Business Programmes, London. p. 92.
14 Tickell, Oliver. 1993. 'Supercar takes high road to the future'. *New Scientist*, 26 June 1993. p. 20.
15 Beard, Jonathan. 1994. 'Green hybrid takes to the track at Le Mans'. *New Scientist*, 26 March 1994.
16 Searle, Graham. 1975. 'Copper in Snowdonia National Park', in Smith, Peter J. (ed.) *The Politics of Physical Resources*. Penguin Education in association with the Open University Press, Harmondsworth.
17 Gregory, Roy. 1971. *The Price of Amenity: Five Studies in Conservation and Government*. Macmillan, London.

17 Greenhouse or Ice Age?

1 See Allaby, Michael. 1990. *Living in the Greenhouse*. Thorsons Publishing Group, London.
2 Pearce, Fred. 1992. 'American sceptic plays down warming fears'. *New Scientist*, 19/26 December 1992. p. 6.
3 Tyndall, John. 1895. *Six Lectures on Light*. Longmans, Green and Co., London. Lecture V.
4 Myers, Norman (ed.). 1985. *The Gaia Atlas of Planet Management*. Pan Books, London. p. 17.
5 Westbroek, Peter, Collins, Matthew J., Jansen, J.H. Fred, and Talbot, Lee M. 1993. 'World archaeology and global change: Did our ancestors ignite the Ice Age?'. *World Archaeology*, **25**, 1, 122. Routledge, London.
6 Eddy, John A. 1977. 'The Case of the Missing Sunspots'. *Scientific American*, May 1977. p. 80.
7 Pennington, Winifred. 1974. *The History of British Vegetation*. 2nd ed. Hodder and Stoughton, London.

18 Nuclear Power – a Faustian Bargain?

1 Wolf, C.P. 1987. 'The NIMBY Syndrome: Its Cause and Cure', in Sterrett, Frances S. (ed.). 1987. *Environmental Sciences*. New York Academy of Sciences, New York.
2 See British Medical Association. 1991. *Hazardous Waste and Human Health*. Oxford University Press, Oxford. pp. 15–18.

3 Wolf, op.cit.
4 Organisation for Economic Co-operation and Development. 1979. *Technology on Trial*. OECD, Paris, p. 50.
5 Lovins, Amory. 1977. *Soft Energy Paths: Toward a Durable Peace*. Penguin Books, Harmondsworth. p. 177.
6 Butler, Declan. 1993. 'Superphénix to become a plutonium "furnace"'. *Nature*, **365**, 381, 30 September 1993.
7 Foley, Gerald. 1976. *The Energy Question*. Penguin Books, Harmondsworth. pp. 175–6.
8 Lovins, Amory. 1974. *Nuclear Power: Technical Bases for Ethical Concern*. Friends of the Earth for Earth Resources Research, London.
9 Thomas, Lewis. 1980. 'The Wonderful Mistake', in *The Medusa and the Snail*. Penguin Books, Harmondsworth.
10 *1990 Britannica Book of the Year*. 1990. Encyclopaedia Britannica. p. 216.
11 Grimston, Malcolm C. 1991. *The Chernobyl Accident: A Review*. AEA Technology internal report.
12 'Chernobyl: Seven Years After'. WHO Press Release WHO/32, 23 April 1993.
13 Ivanov, E.P, Tolochko, G., Lazarev, V.S., and Shuvaeva, L. 1993. 'Child leukaemia after Chernobyl'. *Nature*, **365**, 702, 21 October 1993.
14 From International Chernobyl Project. 1991. *Assessment of Radiological Consequences and Evaluation of Protective Measures*. IAEA, quoted by Grimston, op.cit.
15 Dickman, Steven. 1991. 'Chernobyl effects not as bad as feared'. *Nature*, **351**, 335, 30 May 1991.
16 Grimston, op. cit.
17 Ibid.
18 Forman, David, Cook-Mozaffari, Paula, Darby, Sarah et al. 'Cancer near nuclear installations'. *Nature*, **329**, 499, 8 October 1987.
19 Doll. R., Evans, H.J., and Darby, S.C. 1994. 'Paternal exposure not to blame'. *Nature*, **367**, 678, 24 February 1994.
20 Kenny, Andrew. 1989. 'The Clean Answer to King Coal's Poisonous Reign', in Allaby, Michael (ed.). *Thinking Green*. Barrie and Jenkins, London.
21 At Oklo, Gabon, rock containing more than the usual 0.71 per cent of U-235 became a natural reactor about two billion years ago and continued 'burning' for some 500,000 years until its fuel was depleted. There is clear evidence that most of the by-products of this fission either remained where they formed or moved only a very short distance. In Brazil, a hill called Morro do Ferro is so intensely radioactive that plants growing healthily on it produce an image of themselves when placed on photographic film; radioactive contamination in water draining from the hill is well below the most stringent safety limits. See Allaby, Michael. 1993. *Fire: The Vital Source of Energy*. Facts on File, New York. p. 63.

V RESURRECTING TOMORROW
19 'Prediction is difficult, especially of the future'

1 Calder, Nigel. 1969. *Technopolis*. Macgibbon and Kee, London. pp. 289–90.
2 Toffler, Alvin. 1970. *Future Shock*. The Bodley Head, London. p. 411.
3 Appleyard, Bryan. 1992. *Understanding the Present*. Picador Books, London. p. 237.

20 Migrants from Earth

1 The Planetary Society. 1993. *The Planetary Report*, **XIII**, 5, 21. September/October 19
   The Planetary Society, Pasadena.
2 Campbell, Joseph. 1970. 'The Moon Walk – the Outward Journey', in Campb
   Joseph. 1972. *Myths to Live By*. Souvenir Press, London.
3 Carpenter, Rhys. 1973. *Beyond the Pillars of Hercules*. Universal-Tandem Publishing C
   London. p. 72.
4 Sagan, Carl, Thompson, W. Reid, Carlson, Robert et al. 1993. 'A search for life
   Earth from the Galileo spacecraft'. *Nature*, **365**, 715, 21 October 1993.
5 Allaby, Michael and Lovelock, James. 1984. *The Greening of Mars*. Andre Deuts
   London.
6 McKay, Christopher P., Toon, Owen B., and Kasting, James F. 1991. 'Making M
   habitable'. *Nature*, **352**, 489, 8 August 1991.
7 Hansson, Anders. 1991. *Mars and the Development of Life*. Ellis Horwood, Chichester. p.
   et seq.
8 O'Neill, Bill. 1993. 'Cities in the sky'. *New Scientist*, 2 October 1993. p. 24.
9 *New Scientist*, 10 October 1992. p. 21.
10 Dewdney, A.K. 1992. 'Photovores'. *Scientific American*, September 1992. pp. 18–1
11 Wallich, Paul. 1992. 'Smart wheels'. *Scientific American*, October 1992. p. 12.
12 Concar, David. 1993. 'Can robots come to care for us?' *New Scientist*, 2 October 1993. p.
13 Dyson, Freeman. 1979. *Disturbing the Universe*. Pan Books, London. pp. 232–3.
14 Sagan Carl. 1993. *The Planetary Report*, **XIII**, 4, 23, July/August 1993.

21 Searching for Intelligent Extraterrestrials

1 Shklovskii, I.S. and Sagan, Carl. 1977. *Intelligent Life in the Universe*. Picador Boc
   London. p. 360.
2 Sagan, Carl, Thompson, W. Reid, Carlson, Robert et al. 1993. 'A search for life
   Earth from the Galileo spacecraft'. *Nature*, **365**, 715, 21 October 1993.
3 Shklovskii and Sagan, op. cit. p. 453 et seq.
4 Sagan, Thompson, Carlson et al., op. cit.
5 Henbest, Nigel. 1992. 'SETI: the search continues'. *New Scientist*, 10 October 1992.
   12.
6 Horowitz, Paul. 1993. 'Project META: What Have We Found?' *The Planetary Report*, X
   5, 4, September/October 1993. The Planetary Society, Pasadena.
7 Quoted in Shklovskii and Sagan, op. cit. p. 358.

22 Unravelling Genomes and Designing Species

1 Brown, Terence A. and Brown, Keri A. 1992. 'Ancient DNA and the archaeologi
   Antiquity, **66**, 250, 10 March 1992.

2 Miller, Henry I. 1992. Letter to *Nature*, **360**, 101, 12 November 1992.
3 Kevles, Daniel J. 'Unholy Alliance'. *The Sciences*, September/October 1986. New York Academy of Sciences. p. 25.
4 *Biologist*, **40**, 3, 122, June 1993.
5 Mabberley, D.J. 1987. *The Plant Book*. Cambridge University Press, Cambridge.
6 Crawley, M.J., Hails, R.S., Rees, M. et al. 1993. 'Ecology of transgenic oilseed rape in natural habitats'. *Nature*, **363**, 620, 17 June 1993.
7 Kevles, op.cit.
8 'Brave new world', *Daily Telegraph*, London. 16 August 1993.

23 Science, the Continuing Quest

1 Gombrich, Ernst. 1972. *The Story of Art*. 13th ed. Phaidon Press, London. Opening words of the Introduction.
2 Appleyard, Bryan. 1992. *Understanding the Present*. Picador Books, London. p. 237.
3 Popper, Karl. 1992. *Unended Quest: An Intellectual Autobiography*. Routledge, London. Chapter 16.
4 Appleyard, Bryan. 1993. 'Master of a narrow universe'. *The Independent*, 13 October 1993.
5 Highfield, Roger. 1993. 'Computers hold key to universe says scientist'. *Daily Telegraph*, 27 October 1993.
6 See Penrose, Roger. 1987. 'Minds, Machines and Mathematics', in Blakemore, Colin and Greenfield, Susan (eds.). *Mindwaves*. Basil Blackwell, Oxford.
7 Midgley, Mary. 1992. *Science as Salvation: A Modern Myth and its Meaning*. Routledge, London. p. 65.

24 Science and Ethics

1 Fishel, Simon. 1988. 'Assisted Human Reproduction and Embryonic Surgery: The Ethical Issues', in Callahan, Daniel and Dunstan, G.R. (eds.). *Biomedical Ethics: An Anglo-American Dialogue*, Annals of the New York Academy of Sciences, New York.
2 Macilwain, Colin. 1993. 'Cloning human embryos draws fire from critics'. *Nature*, **365**, 778, 28 October 1993.
3 McLaren, Anne. 1990. 'Biomedical Science: Some Controversial Issues', in Maltoni, Cesare and Selikoff, Irving J. (eds.). *Scientific Issues of the Next Century*. Annals of the New York Academy of Sciences, New York.
4 Midgley, Mary. 1983. *Animals and Why They Matter*. Pelican Book, Harmondsworth. p. 11.
5 Ibid. p. 28.
6 Turner, E.S. 1964. *All Heaven in a Rage*. Michael Joseph, London. p. 202.
7 Quoted by Sechzer, Jeri A. 1983. 'The Ethical Dilemma of Some Classical Animal Experiments', in Sechzer, Jeri A. (ed.). *The Role of Animals in Biomedical Research*, Annals of the New York Academy of Sciences, New York.
8 Quoted by Midgley, op. cit. p. 89.

9 Dawkins, Marian Stamp. 1987. 'Minding and Mattering', in Blakemore, Colin and Greenfield, Susan (eds.). *Mindwaves*. Basil Blackwell, Oxford.
10 Grodsky, Phyllis B. 1983. 'Public Opinion on Animal-based Research: The Unknown Factor in Ethical and Policy Decisions', in Sechzer, op. cit.

25 Reconciling the Two Cultures

1 Macilwain, Colin. 1993. 'SSC decision ends post-war science–government partnership'. *Nature*, **365**, 773, 28 October 1993.
2 Ausubel, Jesse H. 1993. '2020 Vision'. *The Sciences*, November/December 1993. New York Academy of Sciences, New York. p. 15.
3 Snow, C.P. 1993. *The Two Cultures*. Canto Edition, Cambridge University Press, Cambridge.

Bibliography

Alcock, John. 1979. *Animal Behavior: An evolutionary approach.* Sinauer Associates, Sunderland, Massachusetts.

Allaby, Michael (ed.). 1989. *Thinking Green: An Anthology of Essential Ecological Writing.* Barrie and Jenkins, London; 1989. *Guide to Gaia.* Macdonald Optima, London; 1990. *Living in the Greenhouse.* Thorsons Publishing Group, London; 1992. *Water: Its Global Nature.* Facts on File, New York; 1993. *Fire: The Vital Source of Energy.* Facts on File, New York.

Allaby, Michael and Lovelock, James. 1983. *The Great Extinction.* Secker and Warburg, London; 1984. *The Greening of Mars.* Andre Deutsch, London.

Ames, Bruce N. 1992. 'Facts Versus Phantoms'. *Perspectives,* **2**, 3, Winter 1992/3. Wells Associates for Guinness PLC, Lausanne.

Anderson, Christopher. 1992. 'Creationist victory'. *Nature,* **355**, 757. 27 February 1992.

Anon. 1967. *Did Man Get Here by Evolution or by Creation?.* Watch Tower Bible and Tract Society, New York.

Appleyard, Bryan. 1992. *Understanding the Present.* Pan Books, London.

Aristotle. *Politics.* Trans. John Warrington. 1959. Heron Books, London.

Ausubel, Jesse H. 1993. '2020 Vision'. *The Sciences,* November/December 1993. New York Academy of Sciences, New York.

Axelrod, R. 1984. *The Evolution of Cooperation.* Basic Books, New York.

Ayer, A.J. 1956. *The Problem of Knowledge.* Penguin Books, Harmondsworth.

Beard, Jonathan. 1994. 'Green hybrid takes to the track at Le Mans'. *New Scientist,* 26 March 1994.

Blair, Bruce C. and Kendall, Henry W. 1990. 'Accidental Nuclear War'. *Scientific American,* **263**, 6, 19 December 1990.

Blakemore, Colin and Greenfield, Susan (eds.). 1987. *Mindwaves.* Basil Blackwell, Oxford.

Borgstrom, George. 1972. *The Hungry Planet.* 2nd edition. Collier Books, New York.

British Medical Association. 1991. *Hazardous Waste and Human Health.* Oxford University Press, Oxford.

Brown, Terence A. and Brown, Keri A. 1992. 'Ancient DNA and the archaeologist'. *Antiquity,* **66**, 250, 10 March 1992.

BIBLIOGRAPHY

Buderi, Robert. 1990. 'Approval for degrees turned down'. *Nature*, **343**, 501. 8 February 1990.

Burgess, G.H.O. 1967. *The Curious World of Frank Buckland*. John Baker, London.

Burnett, John. 1966. *Plenty and Want: A social history of diet in England from 1815 to the present day.* Nelson, London.

Button, John. 1990. *New Green Pages*. Macdonald Optima, London.

Calder, Nigel. 1969. *Technopolis*. Macgibbon and Kee, London.

Callahan, Daniel and Dunstan, G.R. (eds.). 1988. *Biomedical Ethics: An Anglo-American Dialogue.* Annals of the New York Academy of Sciences, New York.

Campbell, Joseph. 1972. *Myths to Live By*. Viking Press, New York.

Capra, Fritjof. 1975. *The Tao of Physics*. Wildwood House, London; 1982. *The Turning Point.* Wildwood House, London.

Carlyle, Thomas. 1831. 'Diderot', in *Foreign Quarterly Review*, **22**, 1. Paris.

Carpenter, Rhys. 1973. *Beyond the Pillars of Hercules*. Universal-Tandem Publishing Co., London.

Carson, Rachel. 1963. *Silent Spring*. Hamish Hamilton, London.

Charlson, Robert J., Lovelock, James E., Andreae, Meinrat O. and Warren, Stephen G. 1987. 'Oceanic phytoplankton, atmospheric sulphur, cloud albedo and climate'. *Nature*, **326**, 655, 16 April 1987.

Chiao, Raymond Y., Kwiat, Paul G. and Steinberg, Aephraim M. 'Faster than Light?' *Scientific American*, August 1993, p. 38.

Christy, Campbell. 1984. *Nuclear Facts*. Hamlyn, London.

Cloud, Preston. 1970. 'Mined out!', *The Ecologist*, **1**, 2, 25. August 1970.

Cobbett, William. 1830. *Rural Rides*. Republished 1967 by Penguin Books, Harmondsworth.

Cockburn, Andrew. 1991. *An Introduction to Evolutionary Ecology*. Blackwell Scientific Publications, Oxford.

Coghlan, Andy. 1993. 'Nice research, shame about the profits'. *New Scientist*, 14 August 1993. p. 4.

Cole, LaMont C. 1970. 'Can the World Be Saved?' in Johnson, Cecil E. (ed.). *Eco-Crisis.* John Wiley & Sons, New York.

Concar, David. 1993. 'Can robots come to care for us?' *New Scientist*, 2 October 1993.

Crawley, M.J., Hails, R.S., Rees, M., Kohn, D. and Buxton, J. 1993. 'Ecology of transgenic oilseed rape in natural habitats'. *Nature*, **363**, 620, 17 June 1993.

Crichton, Michael. 1991. *Jurassic Park*. Random Century, London.

Crick, Francis and Koch, Christof. 1992. 'The Problem of Consciousness'. *Scientific American*, September 1992.

Cumming, W.P., Hillier, S., Quinn, D.B. and Williams G. 1974. *The Exploration of North America 1630–1776*. Paul Elek, London.

Davenas, E., Beauvais, F., Amara, J., Oberbaum, M., Robinzon, B., Miadonna, A., Tedeschi, A., Pomeranz, B., Fortner, P., Belon, P., Sainte-Laudy, J., Poitevin, B. and Benveniste, J. 1988. 'Human basophil degranulation triggered by very dilute serum against IgE'. *Nature*, **333**, 816, 30 June 1988.

Davies, Paul. 1983. *God and the New Physics*. Dent, London.

Dawkins, Marian Stamp. 1987. 'Minding and Mattering', in Blakemore, Colin and Greenfield, Susan (eds.). *Mindwaves*. Basil Blackwell, Oxford.

De Bell, Garrett (ed.). 1970. *The Environmental Handbook*, Ballantine and Friends of the Earth, New York.

Descartes, René. 1637. *Discourse on the Method*, Book IV. in Anscombe, Elizabeth and Geach, Peter Thomas (eds.). 1970. *Descartes Philosophical Writings*. Nelson's University Paperback, London.

Dewdney, A.K. 1992. 'Photovores'. *Scientific American*, September 1992.

Dickman, Steven. 1991. 'Chernobyl effects not as bad as feared'. *Nature*, **351**, 335, 30 May 1991.

Dirac, Margit W. 1993. Letter to *Scientific American*, August 1993.

Doll, R., Evans, H.J. and Darby, S.C. 1994. 'Paternal exposure not to blame'. *Nature*, **367**, 678, 24 February 1994.

Dubos, René. 1980. *The Wooing of Earth*. The Athlone Press, London.

Dyson, Freeman. 1979. *Disturbing the Universe*. Harper and Row, New York.

Eddy, John A. 1977. 'The Case of the Missing Sunspots'. *Scientific American*, May 1977.

Ehrlich, Paul R. and Ehrlich, Anne H. 1970. *Population, Resources, Environment*. W.H. Freeman, San Francisco; 1970. 'Too Many People', extract from *The Population Bomb*, in De Bell, Garrett (ed.). *The Environmental Handbook*. Ballantine Books, New York.

Fast, Howard (ed.). 1948. *The Selected Works of Tom Paine*. The Bodley Head, London.

Fishel, Simon. 1988. 'Assisted Human Reproduction and Embryonic Surgery: The Ethical Issues', in Callahan, Daniel and Dunstan, G.R. (eds.). *Biomedical Ethics: An Anglo-American Dialogue*. Annals of the New York Academy of Sciences, New York.

Foley, Gerald. 1976. *The Energy Question*. Penguin Books, Harmondsworth.

Forman, David, Cook-Mozaffari, Paula, Darby, Sarah, Davey, Swyneth, Stratton, Irene, Doll, Richard and Pike, Malcolm. 1987. 'Cancer near nuclear installations'. *Nature*, **329**, 499, 8 October 1987.

Franklin, Carl. 1993. 'Did life have a simple start?' *New Scientist*, 2 October 1993.

Fukuyama, Francis. 1992. *The End of History and the Last Man*. Hamish Hamilton, London.

Galbraith, John Kenneth. 1992. *The Culture of Contentment*. Sinclair-Stevenson, London.

Galton, Francis. 1883. *Inquiries into Human Faculty and Its Development*. Everyman edition, J.M. Dent, London.

Gansberg, Alan L. 1983. *Little Caesar: A biography of Edward G. Robinson*. New English Library, Sevenoaks, Kent.

Gleason, Daniel F. and Wellington, Gerard M. 1993. 'Ultraviolet radiation and coral bleaching'. *Nature*, **365**, 836, 28 October 1993.

Gleick, James. 1987. *Chaos: Making a New Science*. Heinemann, London.

Goldsmith, Edward, 1974. 'How to avoid Flixboroughs'. *The Ecologist*, **4**, 6, 202; 1978. *The Stable Society*. The Wadebridge Press, Wadebridge, Cornwall, UK; 1988. *The Great U-Turn: De-industrializing Society*. Green Books, Hartland, Devon, England; 1992. *The Way*. Rider, London.

Goldsmith, Edward, Allen, Robert, Allaby, Michael, Davoll, John and Lawrence, Sam. 1972. 'A Blueprint for Survival'. *The Ecologist*, **2**, 1.

Gombrich, Ernst. 1972. *The Story of Art*. 13th ed. Phaidon Press, London.

Goodall, Jane. 1986. *The Chimpanzees of Gombe: Patterns of Behavior*. The Belknap Press of Harvard University Press. Cambridge, Mass.

Gould, Stephen Jay. 1981. *The Mismeasure of Man*. Penguin Books, Harmondsworth; 1983. *Hen's Teeth and Horse's Toes*. Penguin Books, Harmondsworth.

Gregory, Roy. 1971. *The Price of Amenity: Five Studies in Conservation and Government*. Macmillan, London.

Grimston, Malcolm C. 1991. *The Chernobyl Accident*. AEA Technology.

Grodsky, Phyllis B. 1983. 'Public Opinion on Animal-based Research: The Unknown

Factor in Ethical and Policy Decisions', in Sechzer, Jeri A. (ed.). *The Role of Animals Biomedical Research.* Annals of the New York Academy of Sciences, New York.

Grove, Richard H. 1992. 'Origins of Western Environmentalism'. *Scientific American*, Jul 1992. p. 23.

Groves, Colin P. 1992. 'Drowning in Superstition: Flood Myths and their Significance'. T *Skeptic*, Australian Skeptics, Melbourne; 1991. *A Theory of Human and Primate Evolutio* Clarendon Press, Oxford.

Grylls, Rosalie Glynn. 1953. *William Godwin and his World.* Odhams Press, London.

Hampson, Norman. 1968. *The Enlightenment.* Penguin Books, Harmondsworth.

Hansson, Anders. 1991. *Mars and the Development of Life.* Ellis Horwood, Chichester.

Hawking, Stephen. 1988. *A Brief History of Time.* Bantam Press, London.

Headland, Thomas N. 1990. 'Paradise Revised'. *The Sciences*, September/October. New Yor Academy of Sciences.

Heath-Stubbs, John and Salman, Phillips (eds.). 1984. *Poems of Science.* Penguin Book Harmondsworth.

Heisenberg, Werner. 1971. *Physics and Beyond: Memories of a Life in Science.* George Allen an Unwin, London.

Hemming, John. 1978. *Red Gold: The Conquest of the Brazilian Indians.* Macmillan, London.

Henbest, Nigel. 1992. 'SETI: the search continues'. *New Scientist*, 10 October 1992.

Hey, Tony and Walters, Patrick. 1987. *The Quantum Universe.* Cambridge University Press Cambridge.

Hitching, Francis. 1982. *The Neck of the Giraffe or Where Darwin Went Wrong.* Pan Book London.

Horgan, John. 1993. 'Perpendicular to the Mainstream', a profile of Freeman Dyson *Scientific American*, August 1993; 1993. 'Quantum Philosophy'. *Scientific American*, Jul 1992, p. 77; 1993. 'Profile: Paul Karl Feyerabend'. *Scientific American*, May 1993.

Horowitz, Paul. 1993. 'Project META: What Have We Found?' *The Planetary Report*, **XIII**, 5, September/October 1993. The Planetary Society, Pasadena.

Hovis, R. Corby and Kragh, Helge. 1993. 'P.A.M. Dirac and the Beauty of Physics'. *Scientif American*, May 1993.

Humphrey, Nicholas. 1987. 'The Inner Eye of Consciousness', in Blakemore, Colin an Greenfield, Susan (eds.). *Mindwaves.* Basil Blackwell, Oxford.

Huxley, T.H. 'Man's Relations to Lower Animals' in *Man's Place in Nature and Other Essays.* 190 edition published by J.M. Dent and Sons, London.

Independent Commission on International Development Issues under the Chairman ship of Willy Brandt. 1980. *North-South: A programme for survival.* Pan Books London.

Inglis, Brian. 1971. *Poverty and the Industrial Revolution.* Hodder and Stoughton, London.

Ivanov, E.P, Tolochko, G., Lazarev, V.S. and Shuvaeva, L. 1993. 'Child leukaemia afte Chernobyl'. *Nature*, **365**, 702, 21 October 1993.

Joseph, Lawrence E. 1990. *Gaia: The Growth of an Idea.* Arkana, Penguin Books, London.

Kahn, Hermann, Brown, William and Martel, Leon. 1976. *The Next 200 Years.* Associate Business Programmes, London.

Kapitza, Sergei. 1991. 'Antiscience Trends in the USSR'. *Scientific American*, August 1991

Keay, R.W.J. 1990. 'Presidential Address 1990: A sense of perspective in the tropical rai forest'. *Biologist*, **37**, 3, 73, June 1990.

Kenny, Andrew. 1989. 'The Clean Answer to King Coal's Poisonous Reign', in Allaby Michael (ed.). *Thinking Green.* Barrie and Jenkins, London.

Kevles, Daniel J. 'Unholy Alliance'. *The Sciences*, September/October 1986. New York Academy of Sciences.

King, Mary-Claire. 1993. 'Sexual orientation and the X'. *Nature*, **364**, 288.

Klotz, John W. 1972. *Ecology Crisis*. Concordia Press, London.

Koyré, Alexandre. 1954. Introduction to Anscombe, Elizabeth and Geach, Peter Thomas (eds.). 1970. *Descartes Philosophical Writings*. Nelson's University Paperback, London.

Lewin, Roger. 1989. *Human Evolution: An Illustrated Introduction*. 2nd ed. Blackwell Scientific Publications, Oxford.

Lovelock, J.E. 1979. *Gaia: A new look at life on Earth*. Oxford University. Press, Oxford; 1989. *The Ages of Gaia: A Biography of Our Living Earth*. Oxford University Press, Oxford.

Lovins, Amory. 1974. *Nuclear Power: Technical Bases for Ethical Concern*. Friends of the Earth for Earth Resources Research, London; 1977. *Soft Energy Paths: Toward a Durable Peace*. Penguin Books, Harmondsworth.

Luthi, Theres. 1993. 'If I Were a Rich Man'. *The Sciences*, November/December 1993.

Mabberley, D.J. 1987. *The Plant Book*. Cambridge University Press, Cambridge.

Macilwain, Colin. 1993. 'Cloning human embryos draws fire from critics'. *Nature*, **365**, 778, 28 October 1993; 1993. 'SSC decision ends post-war science – government partnership'. *Nature*, **365**, 773, 28 October 1993.

McKay, Christopher P., Toon, Owen B. and Kasting, James F. 1991. 'Making Mars habitable'. *Nature*, **352**, 489, 8 August 1991.

McLaren, Anne. 1990. 'Biomedical Science: Some Controversial Issues', in Maltoni, Cesare and Selikoff, Irving J. (eds.). *Scientific Issues of the Next Century*. Annals of the New York Academy of Sciences, New York.

Maddox, John, Randi, James, and Stewart, Walter W. 1988. '"High-dilution" experiments a delusion'. *Nature*, **334**, 287, 28 July 1988.

Magee, Bryan. 1982. *Popper*. Fontana Paperbacks, London.

Malthus, Thomas Robert. 1798. *An Essay on the Principle of Population as it Affects the Future Improvement of Society*. Penguin Classics. 1982.

Maltoni, Cesare and Selikoff, Irving J. (eds.). 1990. *Scientific Issues of the Next Century*. Annals of the New York Academy of Sciences, New York.

Marshall, N.B. 1979. *Developments in Deep-Sea Biology*. Blandford Press, Poole, Dorset.

Mason, John Hope. 1979. *The Indispensable Rousseau*. Quartet Books, London.

Matthews, William H., Kellogg, William W. and Robinson, G.D. (eds.). 1971. *Man's Impact on the Climate*. MIT Press, Cambridge, Mass.

Maynard Smith, John. 1979. 'Hypercycles and the origin of life'. *Nature*, **280**, 445, 9 August 1979.

Meadows, Donella H., Meadows, Dennis L., Randers, Jørgen and Behrens, William W. III. 1972. *The Limits to Growth*. Earth Island, London.

Medawar, P.B. and Medawar, J.S. 1978. *The Life Science*. Paladin Books, London.

Mellanby, Kenneth. 1992. *The DDT Story*. British Crop Protection Council, Bracknell, Berkshire.

Mesarovic, Mihaijlo and Pestel, Eduard. 1975. *Mankind at the Turning Point*. Hutchinson, London.

Midgley, Mary. 1983. *Animals and Why They Matter*. Pelican Books, Harmondsworth; 1992. *Science as Salvation: A Modern Myth and its Meaning*. Routledge, London.

Molony, Carol. 1988. 'The Truth About the Tasaday'. *The Sciences*, September/October. New York Academy of Sciences.

More, Sir Thomas. 1516. *Utopia: The First Book of the Communication of Raphael Hythloday*. Heron Books London.

BIBLIOGRAPHY

Mumford, Lewis. 1961. *The City in History*. Penguin Books, Harmondsworth.

Myers, Norman (ed.). 1985. *The Gaia Atlas of Planet Management*. Pan Books, London.

Nicholls, Peter (ed.). 1982. *The Science in Science Fiction*. Michael Joseph, London.

Nowak, Martin A. and Sigmund, Karl. 1992. 'Tit for tat in heterogeneous populations'. *Nature*, **355**, 250, 16 January 1992.

Oldroyd, D.R. 1980. *Darwinian Impacts: An introduction to the Darwinian Revolution*. Open University Press, Milton Keynes.

O'Neill, Bill. 1993. 'Cities in the sky'. *New Scientist*, 2 October 1993.

Organisation for Economic Co-operation and Development. 1979. *Technology on Trial*. OECD, Paris.

Paine, Thomas. 1794. *The Age of Reason: Being an Investigation of True and Fabulous Theology*, in Fast, Howard (ed.). 1948. *The Selected Works of Tom Paine*. The Bodley Head, London.

Pearce, Fred. 1992. 'American sceptic plays down warming fears'. *New Scientist*, 19/26 December 1992; 1991. *Green Warriors*. The Bodley Head, London.

Penrose, Roger. 1987. 'Minds, Machines and Mathematics', in Blakemore, Colin and Greenfield, Susan (eds.). *Mindwaves*. Basil Blackwell, Oxford.

Piel, Gerard. 1992. *Only One World: Our Own to Make and to Keep*. W.H. Freeman, New York.

The Planetary Society. 1993. *The Planetary Report*, **XIII**, 5, 21. September/October 1993. The Planetary Society, Pasadena.

Plato. *The Republic*. Trans John Llewellyn Davies and David James Vaughan. 1919. Macmillan, London.

Polkinghorne, Revd Dr John. 1993. 'Can a scientist pray?', in *RSA Journal*, **CXLI** 5441, July 1993.

Popper, Karl. 1992. *Unended Quest: An Intellectual Autobiography*. Routledge, London.

Rees, Gareth. 1993. 'The longbow's deadly secrets'. *New Scientist*, **138**, 1876, p. 24, 5 June 1993.

Restak, Richard. 1993. 'Brain by Design'. *The Sciences*, September/October 1993. New York Academy of Sciences.

Ridley, Mark. 1985. *The Problems of Evolution*. Oxford University Press, Oxford.

Rose, Hilary and Rose, Steven. 1969. *Science and Society*. Penguin Books, Harmondsworth.

Roszak, Theodore. 1973. *Where the Wasteland Ends: Politics and Transcendence in Post industrial Society*. Faber and Faber, London.

Russell, Bertrand. 1961. *History of Western Philosophy*. 2nd ed. George Allen and Unwin, London.

Russell, Peter. 1982. *The Awakening Earth: The Global Brain*. Ark Paperbacks, London.

Ryder, Richard. 1993. 'Violence and machismo'. *RSA Journal*, **CXLI**, 5443, 706. October 1993.

Ryle, Gilbert. 1949. *The Concept of Mind*. Hutchinson, London.

Sagan, Carl. 1993. *The Planetary Report*, **XIII**, 4, 23. July/August 1993.

Sagan, Carl, Thompson, W. Reid, Carlson, Robert, Gurnett, Donald, and Hord, Charles. 1993. 'A search for life on Earth from the Galileo spacecraft'. *Nature*, **365**, 715, 21 October 1993.

Sahlins, Marshall. 1974. 'The Original Affluent Society'. *The Ecologist*, **4**, 5, 181.

Sale, Kirkpatrick. 1980. *Human Scale*. Coward, McCann and Geoghegan, New York; 1985. *Dwellers in the Land: The Bioregional Vision*. Sierra Club Books, San Francisco.

Schweitzer, Albert. 1947. *The Hospital at Lambaréné During the War Years*. Pub. by The Albert Schweitzer Fellowship of America; 1954. *My Life and Thought*. 2nd ed. Guild Books and George Allen and Unwin, London.

Searle, Graham. 1975. 'Copper in Snowdonia National Park', in Smith, Peter J. (ed.). *The Politics of Physical Resources*. Penguin Education in association with the Open University Press, Harmondsworth.

Sechzer, Jeri A. (ed.). 1983. *The Role of Animals in Biomedical Research*. Annals of the New York Academy of Sciences, New York.

Shamos, Morris H. 1993. 'The 20 Percent Solution'. *The Sciences*, March/April 1993, p. 14. New York Academy of Sciences.

Shelley, Mary. 1818. *Frankenstein or, The Modern Prometheus*. Puffin Books edition. 1989. Penguin Books, London.

Shipley, Joseph T. 1968. *Dictionary of Early English*. Littlefield, Adams and Co., Totowa, New Jersey.

Shklovskii, I.S. and Sagan, Carl. 1977. *Intelligent Life in the Universe*. Picador Books, London.

Slakey, Francis. 1993. 'When the lights of reason go out'. *New Scientist*, 11 September 1993.

Smith, Peter J. (ed.). *The Politics of Physical Resources*. Penguin Education in association with the Open University Press, Harmondsworth.

Snow, C.P. 1993. *The Two Cultures*. Canto Edition, Cambridge University Press, Cambridge.

Tansley, A.G. 1923. *Practical Plant Ecology*. George Allen and Unwin, London.

Terrace, Herbert. 1987. 'Thoughts Without Words' in Blakemore, Colin and Greenfield, Susan (eds.). *Mindwaves*. Basil Blackwell, Oxford.

Theocharis, T. and Psimopoulos, M. 1987. 'Where science has gone wrong'. *Nature*, **329**, 595, 15 October 1987.

Thomas, Keith. 1983. *Man and the Natural World*. Penguin Books, Harmondsworth.

Thomas, Lewis. 1980. 'The Wonderful Mistake', in *The Medusa and the Snail*. Penguin Books, Harmondsworth.

Tickell, Oliver. 1993. 'Supercar takes high road to the future'. *New Scientist*, 26 June 1993.

Titmuss, Richard M. 1987. *The Philosophy of Welfare*. Allen and Unwin, London.

Toffler, Alvin. 1970. *Future Shock*. The Bodley Head, London.

Tranter, Neil. 1973. *Population since the Industrial Revolution: The case of England and Wales*. Croom Helm, London.

Tudge, Colin. 1977. *The Famine Business*. Faber and Faber, London.

Turco, Richard P., Toon, Owen B., Ackerman, Thomas P., Pollack, James B. and Sagan, Carl. 1984. 'The Climatic Effects of Nuclear War'. *Scientific American*, **251**, 2, 23 August 1984.

Turner, E.S. 1964. *All Heaven in a Rage*. Michael Joseph, London.

Twain, Mark. 'The Damned Human Race: I. Was the World Made for Man?' in DeVoto, Bernard (ed.). 1962. *Letters from the Earth*. Fawcett Publications, Greenwich, Conn.

Tyndall, John. 1895. *Six Lectures on Light*. Longmans, Green and Co., London.

US Department of Health, Education, and Welfare. 1969. *Report of the Secretary's Commission on Pesticides and Their Relationship to Environmental Health*. US Government Printing Office, Washington DC.

Veggeberg, Scott. 1993. 'Escape from Biosphere 2'. *New Scientist*, **139**, 1892, 24, 25 September 1993.

Walgate, Robert. 1986. 'Politics before scientific advice'. *Nature*, 322, 762, 28 August 1986.

Wallich, Paul. 1992. 'Smart wheels'. *Scientific American*, October 1992.

Warnock, Mary (ed.). 1962. *Utilitarianism*. Collins/Fontana, London.

Westbroek, Peter. 1992. *Life as a Geological Force: Dynamics of the Earth*. W.W. Norton and Co., New York.

BIBLIOGRAPHY

Westbroek, Peter, Collins, Matthew J., Jansen, J.H. Fred and Talbot, Lee M. 1993. 'World archaeology and global change: Did our ancestors ignite the Ice Age?' *World Archaeology*, **25**, 1, 122. Routledge, London.

White, David Gordon. 1991. *Myths of the Dog-Man*. University of Chicago Press, Chicago.

Wilkins, Adam S. 1993. 'Jurassic hype', letter to *Nature*, **364**, 568, 12 August 1993.

Wolf, C.P. 1987. 'The NIMBY Syndrome: Its Cause and Cure', in Sterrett, Frances S. (ed.). 1987. *Environmental Sciences*. New York Academy of Sciences, New York.

Wolpert, Lewis. 1992. *The Unnatural Nature of Science*. Faber and Faber, London.

The World Commission on Environment and Development. 1987. *Our Common Future*. Oxford University Press.

Young, Arthur. 1808. *General Report on Enclosures*, Appendix III. Republished 1971 by Augustus M. Kelley, New York.

Zohar, Danah and Marshall, Ian. 1994. *The Quantum Society*. Bloomsbury, London.

Index

INDEX

INDEX